An Introduction to Partial Differential Equations (with Maple)

A Concise Course

An Introduction to Partial Differential Equations (with Maple)

A Concise Course

Zhilin Li & Larry Norris
North Carolina State University, USA

NEW JERSEY • LONDON • SINGAPORE • BEIJING • SHANGHAI • HONG KONG • TAIPEI • CHENNAI • TOKYO

Published by

World Scientific Publishing Co. Pte. Ltd.
5 Toh Tuck Link, Singapore 596224
USA office: 27 Warren Street, Suite 401-402, Hackensack, NJ 07601
UK office: 57 Shelton Street, Covent Garden, London WC2H 9HE

Library of Congress Control Number: 2021045205

British Library Cataloguing-in-Publication Data
A catalogue record for this book is available from the British Library.

AN INTRODUCTION TO PARTIAL DIFFERENTIAL EQUATIONS (WITH MAPLE)
A Concise Course

Copyright © 2022 by World Scientific Publishing Co. Pte. Ltd.

All rights reserved. This book, or parts thereof, may not be reproduced in any form or by any means, electronic or mechanical, including photocopying, recording or any information storage and retrieval system now known or to be invented, without written permission from the publisher.

For photocopying of material in this volume, please pay a copying fee through the Copyright Clearance Center, Inc., 222 Rosewood Drive, Danvers, MA 01923, USA. In this case permission to photocopy is not required from the publisher.

ISBN 978-981-122-862-9 (hardcover)
ISBN 978-981-122-863-6 (ebook for institutions)
ISBN 978-981-122-864-3 (ebook for individuals)

For any available supplementary material, please visit
https://www.worldscientific.com/worldscibooks/10.1142/12052#t=suppl

Typeset by Stallion Press
Email: enquiries@stallionpress.com

Preface

The purpose of this book is to provide an introduction to partial differential equations (PDE) for one or two semesters. The book is designed for undergraduate or beginning level graduate students in mathematics, students from physics and engineering, interdisciplinary areas, and others who need to use partial differential equations, Fourier series, Fourier and Laplace transforms. The prerequisite is a basic knowledge of calculus, linear algebra, and ordinary differential equations.

The textbook aims to be practical, elementary, and reasonably rigorous; the book is concise in that it describes fundamental solution techniques for first order, second order, linear partial differential equations for general solutions, fundamental solutions, solution to Cauchy (initial value) problems, and boundary value problems for different PDEs in one and two dimensions, and different coordinates systems. For boundary value problems, solution techniques are based on the Sturm-Liouville eigenvalue problems and series solutions. The book is accompanied with enough well tested Maple files and some Matlab codes that are available online. The use of Maple makes the complicated series solution simple, interactive, and visible. These features distinguish the book from other textbooks available in the related area.

While there are many PDE textbooks around, many of them cover either too much material or are too difficult. We propose to have a practical, elementary, and reasonably rigorous, concise book that describes fundamental solution techniques with the help of Maple.

This is a textbook based on materials that the authors have used in teaching undergraduate courses on partial differential equations at North Carolina State University (NCSU). A web-site

https://zhilin.math.ncsu.edu/PDE_BOOK

has been set up where updated book information including Maple and Matlab files, solution to homework problems, and other related information. This book project was partially supported by NCSU Library Alt-Textbook award (2016-2017). We would also like to thank my students for proofreading the book.

A project to create an open-source online homework system, called WebWork, see https://webwork.maa.org/ for the course of Partial Differential Equations and this book has already started, which will also be updated through the book webpage.

Contents

Preface		**v**
1	**Introduction**	**1**
	1.1 Further reading	7
	1.2 Exercises	7
2	**First order partial differential equations**	**9**
	2.1 Method of changing variables	9
	2.2 Solution to Cauchy problems	11
	2.3 Method of characteristic for advection equations	12
	2.4 Solution of advection equations of boundary value problems . . .	13
	2.5 Boundary value problems of advection equation with $a < 0$ * . . .	15
	2.6 Method of characteristics for general linear first order PDEs . . .	17
	2.7 Solution to first order linear non-homogeneous PDEs with constant coefficients	18
	2.8 Exercises	20
3	**Solution to one-dimensional wave equations**	**23**
	3.1 Solution to Cauchy problems of 1D wave equations	25
	3.2 Normal modes solutions to 1D wave equations of BVPs	29
	3.3 Exercises	31
4	**Orthogonal functions & expansions, and Sturm-Liouville theory**	**33**
	4.1 Orthogonal functions	35
	4.2 Function expansions in terms of orthogonal sets	39

		4.2.1 Approximating functions using an orthogonal set	40
	4.3	Sturm-Liouville eigenvalue problems	42
		4.3.1 Regular and singular Sturm-Liouville eigenvalue problems .	45
	4.4	Theory and applications of Sturm-Liouville eigenvalue problems .	48
	4.5	Application of the S-L eigenvalue theory and orthogonal expansions	50
	4.6	Series solution of 1D heat equations of initial and boundary value problems	53
	4.7	Exercises	56
5	**Various Fourier series, properties and convergence**		**61**
	5.1	Period, piecewise continuous/smooth functions	62
	5.2	The classical Fourier series expansion and partial sums	67
	5.3	Fourier series of functions with arbitrary periods	73
	5.4	Half-range expansions	79
	5.5	Some theoretical results of various Fourier series	83
	5.6	Exercises	87
6	**Series solutions of PDEs of boundary value problems**		**91**
	6.1	One-dimensional wave equations	91
	6.2	Series solution of 1D wave equations with derivative boundary conditions	93
		6.2.1 Summary of series solutions of 1D wave equations with homogeneous linear BC's	95
		6.2.2 Series solution of 1D wave equations of BVPs with a lower order term	96
	6.3	Series solution to 1D heat equations with various BC's	97
		6.3.1 Summary of series solutions of 1D heat equations with homogeneous linear BC's	102
		6.3.2 Steady state solutions of 1D heat equations of BVPs	103
	6.4	Two-dimensional Laplace equations of BVPs on rectangles	105
	6.5	Double series solutions for 2D wave equations of BVPs*	111

	6.6 Method of separation of variables for PDEs of BVPs in polar coordinates	114
	6.7 Series solution of 2D wave equations of BVPs with a radial symmetry	120
	6.8 Series solution to 3D Laplace equations of BVPs with a radial symmetry	123
	6.9 Special functions related to series solutions of partial differential equations of BVPs	127
	6.10 Exercises	129
7	**Fourier and Laplace transforms**	**133**
	7.1 From the Fourier series to Fourier integral representations	133
	7.1.1 Properties of the Fourier transform	137
	7.1.2 The convolution theorem of the Fourier transform	138
	7.2 Use the Fourier transform to solve PDEs	139
	7.3 The Laplace transform	141
	7.3.1 The inverse Laplace transform and the convolution theorem	143
	7.4 Exercises	144
8	**Numerical solution techniques**	**147**
	8.1 Finite difference methods for two-point boundary value problems	150
	8.1.1 Outline of a finite difference method for a two-point BVP	151
	8.1.2 A Matlab code for the model problem	152
	8.2 Finite difference methods for 1D wave equations of BVPs	155
	8.3 Finite difference methods for 1D heat equations	156
	8.4 Finite difference methods for 2D Poisson equations	157
	8.4.1 The matrix-vector form of the FD equations	159
	8.4.2 The SOR(ω) iterative method	161
	8.5 Exercises	162
A	**ODE review and other useful information**	**165**
	A.1 First order ODEs	165

	A.2	Second order linear and homogeneous ODEs with constant coefficients	166
	A.3	Useful trigonometric formulas	167
	A.4	ODE solutions to the Euler's equations	167

B Introduction to Maple — 171

- B.1 The Maple worksheet 172
- B.2 Arithmetic and algebraic operations in Maple 176
  - B.2.1 The assignment operator 178
  - B.2.2 Special symbols in Maple 178
  - B.2.3 Getting started with Maple 179
- B.3 Functions in Maple 179
  - B.3.1 Built in functions in Maple 180
  - B.3.2 Plotting functions in a Maple worksheet 182
  - B.3.3 Animating plots 184
  - B.3.4 Differentiating functions in a Maple worksheet 185
  - B.3.5 Integrating functions in a Maple worksheet 187
  - B.3.6 Functions defined on integers 188
  - B.3.7 Partial sums 188
  - B.3.8 Maple's repetition statement 191
- B.4 Computing sine, cosine, and full Fourier series 192
  - B.4.1 Fourier sine series 192
  - B.4.2 Fourier cosine series 193
  - B.4.3 Fourier series 195
- B.5 Families of orthogonal functions in Maple 197
  - B.5.1 The family of sine functions $\{\sin(n\pi x/L)\}_{n=1}^{\infty}$ 197
  - B.5.2 The family of functions $\{1, \cos(n\pi x/L)\}_{n=1}^{\infty}$ 197
  - B.5.3 The family of functions $\{1, \cos(n\pi x/L), \sin(n\pi x/L)\}_{n=1}^{\infty}$ 197
  - B.5.4 The Legendre polynomials 198

	B.5.5	The Bessel functions	198
B.6		Quick guide for simple use of Maple	199

Bibliography   201

Index   203

Chapter 1

Introduction

A *differential equation* involves derivatives of an unknown function of one independent variable (say $u(x)$), or partial derivatives of an unknown function of more than one independent variable (say $u(x, y)$, or $u(t, x)$, or $u(t, x, y, z)$ etc.). Differential equations have been used extensively to model many problems in daily life, in fluid and solid mechanics, biology, material sciences, economics, ecology, sports and computer sciences.[1] Examples include the Laplace equation for potentials, the Navier-Stokes equations in fluid dynamics, biharmonic equations for stresses in solid mechanics, and the Maxwell equations in electro-magnetics.

The main part of this textbook is to learn different linear partial differential equations and some techniques to find their solutions. Solutions to differential equations often have physical meanings such as temperature, velocity and acceleration fields, concentration, populations, trajectories of moving objects, stock price etc. With solutions of some differential equations, for an example, we can compute the drag, lift, and resistance force of a flying airplane. We can use solutions of differential equations to *predict* or compute many physical quantities and use the information to *design* or *control* solutions for practical applications. Often, better understanding of differential equations is essential to *improve mathematical models*. Another part of this textbook is about Fourier series and analysis that have practical applications in wave propagation, radio or television broadcasting, and fast computing based on fast Fourier transforms (FFT), and in solving partial differential equations using series solutions.

However, although differential equations have wide applications, not many can be solved exactly in terms of elementary functions such as polynomials, $\log x$, e^x, trigonometric functions ($\sin x$, $\cos x$, ...) etc., and their combinations. Even if a differential equation can be solved analytically, considerable effort and sound

---

[1] There are other models in practice, for example, statistical models.

mathematical theory are often needed, and the closed form of the solution may be too messy to be useful. If the analytic solution of the differential equation is unavailable or too difficult to obtain, or takes some complicated form that is unhelpful to use, we may try to find an approximate solution using two different approaches

- Semi-analytic methods. Sometimes we can use series, integral equations, perturbation techniques, or asymptotic methods to obtain approximate solutions to differential equations.

- Numerical solution methods. The rapid development in modern computers has provided another powerful tool in solving differential equations, called numerical solutions of differential equations. Nowadays, many applications such as weather forecasts, space shuttles lunches, robots, heavily depend on super computer simulations. There are tons of books, software packages, numerical methods, online classes for solving differential equations numerically, which is a developing area of study and research and provides an effective way in solving many problems that were impossible to solve before.

In this book, we mainly adopt the first approach and focus on either analytic solutions or series solutions.

If a differential equation whose solution has only one independent variable, then the differential equation is called an *ordinary differential equation* (ODE). We should have seen many ODE examples before. Below are two simple examples,

$$\frac{dy}{dx} = x; \qquad \frac{dy}{dx} = y.$$

The solutions to the above ODEs are $y(x) = \frac{x^2}{2} + C$ and $y(x) = Ce^x$, respectively, for arbitrary constant C, which means that if we plug the solution into the differential equation, we will get an identity between the left and hand right hand sides of the differential equation.

If a differential equation whose solution has more than one independent variables, then the differential equation is called a *partial differential equation* (PDE). We use the partial derivative symbol $\frac{\partial}{\partial}$ to represent a partial derivative with one particular (independent) variable such as $\frac{\partial u}{\partial x}$, $\frac{\partial u}{\partial t}$ etc. Below are some examples.

Example 1.1. *Solve the following partial differential equation,*

$$\frac{\partial u}{\partial x} = x \quad or \quad \frac{\partial u}{\partial x}(x,t) = x. \qquad (1.1)$$

Sometimes, we can specify the independent variables in the PDE as in the second expression above to avoid possible confusions.

Solution: In the above PDE, since there is only one derivative with respect to x, we can treat the PDE as an ordinary differential equation while regarding the second variable as a parameter (constant) to get $u(x,t) = \frac{x^2}{2} + C(t)$ for any differentiable function $C(t)$. Now we can check that $u(x,t) = \frac{x^2}{2} + C(t)$ is indeed a solution to the PDE. To to so, first we differentiate $u(x,t)$ with respect to x to get $\frac{\partial u}{\partial x} = x + 0$ and plug it into the PDE to have

$$\text{the left hand side} = \frac{\partial u}{\partial x} = x + 0 = \text{the right hand side}.$$

Thus, we have verified that $u(x,t) = \frac{x^2}{2} + C(t)$ is a solution to the PDE for arbitrary differentiable function $C(t)$.

Example 1.2. *Check that $u(x,t) = f(x-at)$ is a solution to the partial differential equation,*

$$\frac{\partial u}{\partial t} + a\frac{\partial u}{\partial x} = 0, \tag{1.2}$$

where a is a constant and $f(s)$ is an arbitrary differentiable function.

Solution: First we differentiate $u(x,t)$ with respect to x using the chain rule to get $\frac{\partial u}{\partial x} = f'(x-at)$. Note that since $f(s)$ is a function of one variable, we can use the symbol f'. Similarly, we differentiate $u(x,t)$ with respect to t using the chain rule to get $\frac{\partial u}{\partial t} = f'(x-at)(-a)$. We plug the partial derivatives $\frac{\partial u}{\partial x} = f'(x-at)$ and $\frac{\partial u}{\partial t} = f'(x-at)(-a)$ into the PDE to get

$$LHS = \frac{\partial u}{\partial t} + a\frac{\partial u}{\partial t} = f'(x-at)(-a) + af'(x-at) = 0 = RHS.$$

Note that LHS and RHS stand for the left hand side and the right hand side, respectively. Thus, we have verified that $u(x,t) = f(x-at)$ is a solution to the partial differential equation.

Example 1.3. *Check that $u(x,y) = x^2 + y^2 + C_1 x + C_2 y + C_3$ is a solution to the partial differential equation,*

$$\frac{\partial^2 u}{\partial x^2} + \frac{\partial^2 u}{\partial y^2} = 4, \tag{1.3}$$

where, C_1, C_2, and C_3 are constants. We should pay attention to the high order differential notations.

Solution: First we differentiate $u(x,y)$ with respect to x and y twice, respectively to have

$$\frac{\partial u}{\partial x} = 2x + C_1, \quad \frac{\partial}{\partial x}\frac{\partial u}{\partial x} = \frac{\partial^2 u}{\partial x^2} = 2,$$

$$\frac{\partial u}{\partial y} = 2y + C_2, \quad \frac{\partial}{\partial y}\frac{\partial u}{\partial y} = \frac{\partial^2 u}{\partial y^2} = 2.$$

We plug them into the PDE to get

$$LHS = \frac{\partial^2 u}{\partial x^2} + \frac{\partial^2 u}{\partial y^2} = 2 + 2 = 4 = RHS.$$

Thus, we have verified that $u(x,y) = x^2 + y^2 + C_1 x + C_2 y + C_3$ is a solution to the partial differential equation.

Note that for ordinary differential equations, the solution can differ by a constant while for partial differential equations, the solution can differ by *functions*. Some examples of ODE/PDE are listed below.

1. Initial value problems (IVP). The canonical form of a first order system is

$$\frac{d\mathbf{y}}{dt} = \mathbf{f}(t, \mathbf{y}), \quad \mathbf{y}(t_0) = \mathbf{y}_0. \tag{1.4}$$

A higher order ordinary differential equation of one variable can be rewritten as a first order system. For example, a second order ordinary differential equation

$$u''(t) + a(t)u'(t) + b(t)u(t) = f(t),$$
$$u(0) = u_0, \quad u'(0) = v_0, \tag{1.5}$$

can be converted into a first order system by setting $y_1(t) = u$ and $y_2(t) = u'(t)$ with $y_1(0) = u_0$ and $y_2(0) = v_0$. Note that the two conditions that uniquely determine the solution to the differential equations are all defined at $t = 0$, a distinguished feature of an initial value problem.

2. Boundary value problems (BVP). Below are two examples of an ODE BVP. The first one is one-dimensional,

$$u''(x) + a(x)u'(x) + b(x)u(x) = f(x),$$
$$u(0) = u_0, \quad u(1) = u_1. \tag{1.6}$$

Note that the two conditions above are defined at different points ($x = 0$ and $x = 1$). The second example is a BVP example of a partial differential equation (PDE) in two dimensions,

$$\begin{aligned} u_{xx} + u_{yy} &= f(x,y), \quad (x,y) \in \Omega, \\ u(x,y) &= u_0(x,y), \quad (x,y) \in \partial\Omega, \end{aligned} \quad (1.7)$$

in a domain Ω with boundary $\partial\Omega$, where

$$u_x = \frac{\partial u}{\partial x}, \quad u_{yy} = \frac{\partial^2 u}{\partial y^2} \quad (1.8)$$

and so on for simplicity of notations if there are no confusions occur. The above PDE is linear and classified as *elliptic*. There are two other classifications for linear PDE, namely, *parabolic* and *hyperbolic*, which will be briefly discussed later in this section. The PDE above is called a two-dimensional (2D) Poisson equation. If $f(x,y) = 0$, it is a two-dimensional Laplace equation.

3. Boundary and initial value problems, *e.g.*,

$$\begin{aligned} u_t &= c^2 u_{xx} + f(x,t), \quad 0 < x < 1, \\ u(0,t) &= g_1(t), \quad u(1,t) = g_2(t), \quad \text{BC}, \\ u(x,0) &= u_0(x), \quad \text{IC}, \end{aligned} \quad (1.9)$$

where BC stands for boundary condition(s) while IC for initial condition(s). We call $f(x,t)$ a source term. If $f(x,t) = 0$, the PDE is called a one-dimensional (1D) heat equation, which is a parabolic PDE. Note that the PDE $u_t = -c^2 u_{xx}$ is called a backward heat equation. A nonzero perturbation at some time instances will result an exponential growth in the solution as t increases. A two-dimensional heat equation has the following form

$$u_t = c^2 \left(u_{xx} + u_{yy} \right). \quad (1.10)$$

4. Eigenvalue problems, *e.g.*,

$$\begin{aligned} u''(x) &= \lambda u(x), \\ u(0) &= 0, \quad u(1) = 0. \end{aligned} \quad (1.11)$$

In this example, both the function $u(x)$ (the *eigenfunction*) and the scalar λ (the *eigenvalue*) are unknowns.

5. Diffusion and reaction equations, e.g.,

$$\frac{\partial u}{\partial t} = \nabla \cdot (\beta \nabla u) + \mathbf{a} \cdot \nabla u + f(u) \tag{1.12}$$

where $\mathbf{a}$ is a constant vector, ∇ is the gradient operator which is the derivative $\nabla u(x) = \frac{du}{dx}$ in 1D, and $\nabla u(x,y) = [\frac{\partial u}{\partial x}, \frac{\partial u}{\partial y}]^T$ in 2D, $\nabla \cdot (\beta \nabla u)$ is called a diffusion term, $\mathbf{a} \cdot \nabla u$ is called an advection term, and $f(u)$ is called a reaction term.

6. Wave equations in 1D have the following form

$$u_{tt} = c^2 u_{xx}, \tag{1.13}$$

where $c > 0$ is called the wave speed. The PDE is hyperbolic. 2D wave equations have the general form

$$u_{tt} = c^2 \left(u_{xx} + u_{yy} \right). \tag{1.14}$$

7. Systems of PDEs. The incompressible Navier-Stokes model is an important nonlinear example for modeling incompressible flows:

$$\begin{aligned} \rho \left(\mathbf{u}_t + (\mathbf{u} \cdot \nabla) \mathbf{u} \right) &= \nabla p + \mu \Delta \mathbf{u} + \mathbf{F}, \\ \nabla \cdot \mathbf{u} &= 0. \end{aligned} \tag{1.15}$$

which has three equations in 2D, and four equations in 3D.

In this book, we will consider *linear* PDEs mostly in one dimension (1D) or two dimensions (2D). A 2D linear PDE has the following general form

$$\begin{aligned} & a(x,y)u_{xx} + 2b(x,y)u_{xy} + c(x,y)u_{yy} \\ & + d(x,y)u_x + e(x,y)u_y + g(x,y)u(x,y) = f(x,y), \end{aligned} \tag{1.16}$$

where the coefficients are independent of $u(x,y)$ so the equation is linear in u and its partial derivatives. In the example above, the solution of the 2D linear PDE is sought in some bounded domain Ω. According to the behaviors of the solutions, the PDE (1.16) is classified as the following three categories:

- Elliptic if $b^2 - ac < 0$ for all $(x,y) \in \Omega$,

- Parabolic if $b^2 - ac = 0$ for all $(x,y) \in \Omega$, and

- Hyperbolic if $b^2 - ac > 0$ for all $(x,y) \in \Omega$.

For some well-known PDEs, for examples, heat equations are parabolic; advection and wave equations are hyperbolic; Laplace and Poisson equations are elliptic. Appropriate solution methods typically depend on the equation class.

For a first order system

$$\frac{\partial \mathbf{u}}{\partial t} = A(\mathbf{x})\frac{\partial \mathbf{u}}{\partial \mathbf{x}}, \quad (1.17)$$

the classification is determined from the eigenvalues of the coefficient matrix $A(\mathbf{x})$. The system if hyperbolic if all eigenvalues are real; otherwise it can be elliptic or parabolic.

1.1 Further reading

This textbook provides an introduction to differential equations and Fourier analysis. There are many textbooks on this topic. Each textbook has its own characteristics. Some are long and comprehensive; some are more theoretical, and some are problem solving orientated. At the Department of Mathematics, North Carolina State University, the following textbooks have been used by different instructors (an incomplete list).

- Partial Differential Equations with Fourier Series and Boundary Value Problems by Nakhlé H. Asmar [1].

- Applied Partial Differential Equations (Undergraduate Texts in Mathematics) by David J. Logan, [9]

- Introduction to Applied Partial Differential Equations by John M. Davis [2].

Advanced partial differential equations can be found in [3, 6] and many others. We would also recommend students to Schaum's outline series for summaries, applications, solved problems, and practices, [12, 13]. Often the solutions to partial differential equations are complicated especially with series solutions. It is beneficial to use some powerful packages such as Maple [4] or Mathematics for symbolic derivations and visualizations, and Matlab [5] for computations and visualizations. In terms of numerical solution techniques to PDEs, we refer the readers to [7, 8, 10, 14, 15].

1.2 Exercises

E1.1 ODE Review: Find general solutions or solutions to the following problems.

(a) $y'(x) + y(x) = 1$.

(b) $y'(x) = -\frac{y(x)}{2}$.

(c) $y'(x) = -\frac{x}{2}$, $y(2) = 3$.

(d) $y'(x) - 2y(x) = \sin x$.

(e) $y'(x) + xy(x) = x$, $y(0) = 0$.

(f) $xdy = ydx$.

E1.2 The hyperbolic sine and cosine functions are defined as
$$\sinh x = \frac{e^x - e^{-x}}{2}, \quad \cosh x = \frac{e^x + e^{-x}}{2}.$$

(a) Check they are solutions to ODE $y'' - y = 0$.

(b) Express e^x and e^{-x} in terms of hyperbolic sine and cosine functions.

(c) Find the Wronskian of $W(\sinh x, \cosh x)$. Can it be zero?

E1.3 Find the general solution of the following partial differential equations assuming that the solution is $u(x,t)$.

(a) $\dfrac{\partial u}{\partial x} = 0$.

(b) $\dfrac{\partial u}{\partial t} = 0$.

(c) $\dfrac{\partial u}{\partial t} = f(x)$, where $f(x)$ is a given function.

(d) $\dfrac{\partial^2 u}{\partial x \partial t} = 0$.

E1.4 Verify that

(a) $u(x,y) = x^2 + y^2$ satisfies the Poisson equation $u_{xx} + u_{yy} = 4$.

(b) $u(x,y) = \log \sqrt{x^2 + y^2}$ satisfies the 2D Laplace equation $u_{xx} + u_{yy} = 0$ if $x^2 + y^2 \neq 0$. **Hint:** You can use Maple to verify.

E1.5 Verify that a solution to the heat equation $u_t = ku_{xx}$ is given by $u(x,t) = \dfrac{1}{\sqrt{4\pi kt}} e^{-x^2/(4kt)}$. It is called the fundamental solution of the 1D heat equation. **Hint:** Maple can be used.

E1.6 Show that $u(r,\theta) = \log r$ and $u(r,\theta) = r \cos \theta$ are both solutions to the two-dimensional Laplace equation in the polar coordinates,
$$u_{rr} + \frac{1}{r}u_r + \frac{1}{r^2}u_{\theta\theta} = 0.$$

Chapter 2

First order partial differential equations

One of the simplest first order partial differential equation (PDE) may be the advection equation

$$\frac{\partial u}{\partial t} + a\frac{\partial u}{\partial x} = 0, \quad \text{or} \quad u_t + au_x = 0, \tag{2.1}$$

where a is constant at this moment, t and x are independent variables, $u(x,t)$ is the dependent variable that to be solved. In most of applications, t often stands for the time, and x stands for the space, and a is called a wave speed. The PDE is called a one-dimensional, first order, linear, constant coefficient, and homogeneous one. Although there are two independent variables, it is called one-dimensional (1D) advection equation since there is only one space variable x. The PDE is classified as a hyperbolic one, and it is also called a one-way wave equation, or a transport equation.

2.1 Method of changing variables

There are several ways to find general solutions of an advection partial differential equation. One of them is the method of changing variables. The idea is to change the partial differential equation to an ordinary differential equation (ODE) so that we can use an ODE solution method to solve the problem. A simplest way of changing variables is the following,

$$\begin{cases} \xi = x - at, \\ \eta = t, \end{cases} \quad \text{or} \quad \begin{cases} x = \xi + a\eta, \\ t = \eta. \end{cases} \tag{2.2}$$

Under such a transform, we have $u(x,t) = u(\xi + a\eta, \eta)$ denoted as $U(\xi, \eta) = u(\xi + a\eta, \eta)$. Then, we represent the original PDE in terms of the new variables using the chain rule to have

$$\frac{\partial u}{\partial t} = \frac{\partial U}{\partial \xi}\frac{\partial \xi}{\partial t} + \frac{\partial U}{\partial \eta}\frac{\partial \eta}{\partial t} = -a\frac{\partial U}{\partial \xi} + \frac{\partial U}{\partial \eta},$$

$$\frac{\partial u}{\partial x} = \frac{\partial U}{\partial \xi}\frac{\partial \xi}{\partial x} + \frac{\partial U}{\partial \eta}\frac{\partial \eta}{\partial x} = \frac{\partial U}{\partial \xi} + \frac{\partial U}{\partial \eta}.$$

Plug them into the original PDE (2.1), we get

$$\frac{\partial U}{\partial \eta} = 0. \tag{2.3}$$

Integrating both sides above with respect to η, we get $U(\xi, \eta) = C$. Note that in an ODE, C is an arbitrary constant. But in a PDE, it can be arbitrary differential function of ξ, denoted as $f(\xi)$. Thus we get the solution

$$u(x, t) = u(\xi + a\eta, \eta) = U(\xi, \eta) = f(\xi) = f(x - at). \tag{2.4}$$

It is straightforward to check that $u(x, y)$ above is indeed a solution to the PDE (2.1). It is called the *general solution* of the PDE since there is no condition attached to the problem. Note that $u(x, t) = f(x - at) = f(a(x/a - t)) = F(x/a - t)$ and the general solutions can have different expressions that are essentially the same.

> **The General Solution of 1D Advection Equation** $u_t + au_x = 0$ is $u(x, t) = f(x - at)$ **for any differentiable function** $f(x)$.

Example 2.1. *The general solution to* $2\frac{\partial u}{\partial t} - 3\frac{\partial u}{\partial x} = 0$ *is* $u(x, t) = F(x + \frac{3}{2}t) = 0$ *or* $u(x, t) = F(2x + 3t)$ *for any differentiable function* $F(x)$.

Remark 2.1. *We require $f(x)$ to be differentiable so that $u(x, t)$ satisfies the PDE at every point (x, t). Such a solution is called a classical or strong solution of the PDE. In many applications, however, a function satisfies the PDE almost everywhere but at a few isolated points or lines or surfaces where the solution maybe discontinuous. Such a solution is called a* weak solution.

Note that there are more than one ways of changing variables. In general, we can use

$$\begin{cases} \xi = a_{11}x + a_{12}t, \\ \eta = a_{21}x + a_{22}t, \end{cases} \tag{2.5}$$

where a_{ij}'s are parameters of a transformation matrix $A = \{a_{ij}\}$ that satisfies $det(A) \neq 0$. We can choose a_{ij}'s so that the PDE in terms of the new variables is simple, like an ODE, so that we can solve it easily. In the discussion above we have $a_{11} = 1$, $a_{12} = -a$, $a_{21} = 0$, and $a_{22} = 1$.

2.2 Solution to Cauchy problems

A Cauchy problem is an initial value problem that is defined in the entire space with an initial condition, that is

$$\frac{\partial u}{\partial t} + a\frac{\partial u}{\partial x} = 0, \quad -\infty < x < \infty, \tag{2.6}$$

$$u(x,0) = u_0(x), \tag{2.7}$$

where $u_0(x)$ is a function defined in $(-\infty, \infty)$. Since we know that the general solution is $u(x,t) = f(x - at)$, we have $u(x,0) = f(x) = u_0(x)$. Thus the solution to the Cauchy problem is

$$u(x,t) = u_0(x - at), \tag{2.8}$$

where $u_0(x)$ is called an initial condition. The solution $u(x,t) = u_0(x - at)$ means that the solution at (x,t) is the same as the initial solution at $(x - at, 0)$. When $a > 0$, $x - at < x$, the solution propagates towards right without changing the shape. That is why it is called a one-way wave equation, or advection equation.

The Solution to a Cauchy Problem of an Advection Equation $u_t + au_x = 0$, $-\infty < x < \infty$, $u(x,0) = g(x)$ **is**

$$u(x,t) = g(x - at) \tag{2.9}$$

for a given differentiable function $g(x)$.

Example 2.2. Let $a = 2$ and

$$u_0(x) = \begin{cases} \sin x & -\pi \leq x \leq \pi, \\ 0 & otherwise. \end{cases}$$

The solution to the Cauchy problem is

$$u_0(x,t) = \begin{cases} \sin(x-2t) & -\pi \leq x - 2t \leq \pi \\ 0 & \text{otherwise.} \end{cases}$$

In Figure 2.1, we plot the solution at $t = 0$ and $t = 2.5$, we can see that the solution is simply shifted to the right.

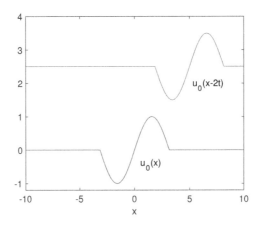

Figure 2.1. *Plot of the initial condition $u_0(x)$ and the solution $u(x,t)$ to the advection equation at $t = 2.5$ with the wave speed $a = 2$.*

2.3 Method of characteristic for advection equations

A characteristic to a partial differential equation is a set in which the solution to the PDE is a constant (does not change). For a first order PDE of the form $u_t + p(x,t)u_x = f(x,t)$, a characteristic is often a continuous curve $(t(s), x(s))$ with a parameter of s, for example, the arc-length of the curve. Let us examine an advection equation $u_t + au_x = 0$ first. Since along the characteristic, the solution $u(x,t) = C$ is a constant, we differentiate the equation on both sides with respect to t to get

$$\frac{\partial u}{\partial t} + \frac{\partial u}{\partial x}\frac{dx}{dt} = 0.$$

Since $u(x,t)$ is the solution to the PDE, we have to have $\frac{dx}{dt} = a$ or $x = at + \bar{C}$. Thus we have $\bar{C} = x - at$. Since $u(x,t)$ is a constant along the line (the characteristic), we have

$$u(x,t) = u(\bar{C}, 0) = u_0(\bar{C}) = u_0(x - at), \qquad (2.10)$$

where $u_0(x)$ is the initial condition. Often we can simply write $\bar{C} = C$. Note that in a partial differential equation, an arbitrary constant often corresponds to an arbitrary function, so we have $C = f(x - at) = u(x,t)$. Once again, we get the general solution using a different method. It is important to know that along a curve $\bar{x} - x = a(\bar{t} - t)$, the solution $u(x,t)$ is a constant, which is the basis to determine appropriate boundary conditions for boundary value problems.

2.4 Solution of advection equations of boundary value problems

Now consider an initial and boundary value problem of an advection equation,

$$\frac{\partial u}{\partial t} + a\frac{\partial u}{\partial x} = 0, \qquad 0 < x < L, \tag{2.11}$$

$$u(x,0) = u_0(x), \qquad 0 < x < L, \tag{2.12}$$

for a positive constant L. We need one or two boundary conditions to make the problem well-posed, that is, the conditions that make the solution exist and unique. Given a point (x,t), $0 < x < L$ and $t > 0$, we can use the method of characteristic to track back the solution to either the initial condition or the boundary condition whichever is the first hit by the characteristic in the domain.

For example, assume that $a > 0$, see the left diagram in Figure 2.2 for an illustration. The line $x = at$ passes through the origin and divide the domain, a strip in the first quadrant, as two parts.

Solution in the lower right triangle: In this domain, we should have one of the following, $x < at$ or $x > at$. Which one is it? Usually we can select a point to decide. At the point $x = L/2$, $t = 0$, we have $L/2 > a \cdot 0 = 0$, which means $x > at$. Thus, the domain is characterized as $x > at$. Next, we trace back the solution $u(x,t)$ to the initial condition, not that the boundary condition, why? To do so, we temporarily fix a point (x,t), and write down the characteristic line using (x,t) and the slope a in the x-t plane, or $1/a$ in the t-x plane,

$$(X - x) = a(T - t), \tag{2.13}$$

where (x,t) is a point that we want to find the solution of $u(x,t)$, and (X,T) is any point on the straight line. If the line intersection the x-axis, that is, $T = 0$ for some X^* between zero and L, then the solution is determined from the initial condition. By setting $T = 0$, we get $X^* = x - at$. Thus we have

$$u(x,t) = u(X^*, 0) = u_0(X^*) = u_0(x - at), \tag{2.14}$$

which is the same as the solution to the Cauchy problem if $0 < at < x < L$ when $a > 0$.

Solution in the strip above the right triangle part: If $0 < x < at$, then the intersection of the line (characteristic) and the x-axis is $X^* = x - at < 0$ that is out of the solution domain. The line (characteristic) also intersects the t axis at $X = 0$ for some T^*. Thus, we set $X = 0$ to solve for the T to get $-x = a(T^* - t)$, or $T^* = t - x/a$. Thus, the solution is from a boundary condition, say $g(0,t) = g_l(t)$,

$$u(x,t) = u(0, T^*) = g_l(T^*) = g_l\left(t - \frac{x}{a}\right). \tag{2.15}$$

In summary, for $a > 0$, we need to prescribed a boundary condition at $x = 0$, say, $u(0,t) = g_l(t)$, here $g_l(t)$ means the boundary condition at the left end.

Solution to an Advection Equation of BVP:

$$u_t + au_x = 0, \quad a > 0, \quad 0 < x < L,$$
$$u(x,0) = u_0(x), \qquad u(0,t) = g_l(t) \tag{2.16}$$

is $u(x,t) = \begin{cases} u_0(x - at) & 0 < at < x < L,\ t > 0 \\ g_l\left(t - \dfrac{x}{a}\right) & 0 < x < \min\{at, L\} \text{ and } t > 0. \end{cases}$

Example 2.3. *Solve the boundary value problem:*

$$2u_t + 3u_x = 0, \quad 0 < x < 3,$$
$$u_0(x) = \sin(5\pi x), \quad 0 < x < 3, \qquad u(0,t) = \sin(3t).$$

First we suggest to write the PDE in the standard form $u_t + \frac{3}{2}u_x = 0$. From the formula above, we get the following solution to the BVP,

$$u(x,t) = \begin{cases} \sin\left(5\pi(x - \dfrac{3t}{2})\right) & 0 < \dfrac{3t}{2} < x < 3,\ t > 0, \\ \sin\left(3(t - \dfrac{2x}{3})\right) & 0 < x < \min\left\{\dfrac{3t}{2}, 3\right\} \text{ and } t > 0. \end{cases}$$

2.5 Boundary value problems of advection equation with $a < 0$ *

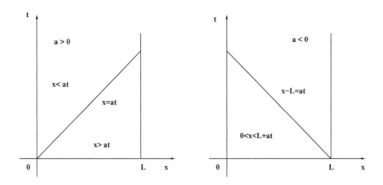

Figure 2.2. *Diagrams of the regions where the solution of an advection is determined either by an initial or a boundary condition. The left diagram is for $a > 0$ while the right is for $a < 0$.*

2.5 Boundary value problems of advection equation with $a < 0$ *

With similar discussions, we know that the initial and boundary value problem,

$$\frac{\partial u}{\partial t} + a\frac{\partial u}{\partial x} = 0, \quad 0 < x < L, \tag{2.17}$$

$$u(x, 0) = u_0(x), \quad 0 < x < L, \tag{2.18}$$

with $a < 0$ requires a boundary condition at $x = L$, say, $u(L, t) = g_r(t)$, which needs to be specified, as illustrated in the right diagram in Figure 2.2.

The line equation $x = at$ will be out of the solution domain, which is useless anymore. We should use a line equation like $x = at + C$ that can cut both the axis $t = 0$ (for initial condition) and the boundary $x = L$. Obviously, the line equation $x = at + L$ passes through $(L, 0)$ and divides the domain in the first quadrant as two parts; in one region we have $0 < x < at + L$; in the other region we have $L + at < x < L$. Once again, we can take a point $(L/2, 0)$ to check. Since $L/2 < L$, the triangle region is described by $x < at + L$.

Solution in the lower left triangle: In this domain, we have $0 < x < at+L$. Next, we trace back the solution $u(x, t)$ to the initial condition, not the boundary condition, why? To do so, we write down the characteristic line using a point (x, t) and the slope a in the x-t plane, or $1/a$ in the t-x plane,

$$(X - x) = a(T - t), \tag{2.19}$$

where (x, t) is a point that we want to find the solution of $u(x, t)$, (X, T) is any point on the straight line. If the line intersection the x-axis, that is, $T = 0$ for some X^* between zero and L, then the solution is determined from the initial condition. By setting $T = 0$, we get $X^* = x - at$. Thus we have

$$u(x,t) = u(X^*, 0) = u_0(X^*) = u_0(x - at), \tag{2.20}$$

which is the same as the solution to the Cauchy problem if $0 < x < at + L$.

Solution in the strip above the right triangle: If $at + L < x < L$, the intersection of the line (characteristic) and the x-axis is $X^* = x - at > L$ or $X^* > at + L > L$ that is out of the solution domain. The line (characteristic) also intersects the line $X = L$ for some T^*. Thus, we set $X = L$ to solve for the T to get $L - x = a(T^* - t)$, or $T^* = t + (L - x)/a$ and the solution is from the boundary condition

$$u(x,t) = u\left(L, t + \frac{L-x}{a}\right) = g_r\left(t + \frac{L-x}{a}\right) \quad \text{if } L + at < x < L.$$

In summary, the solution when $a < 0$ is

$$u_0(x,t) = \begin{cases} u_0(x - at) & 0 < x < L + at; \; t > 0, \\ g_r\left(t + \dfrac{L-x}{a}\right) & \max\{0, L + at\} < x < L, \; t > 0. \end{cases} \tag{2.21}$$

Example 2.4. *Given the boundary value problem below*

$$\frac{\partial u}{\partial t} - 2\frac{\partial u}{\partial x} = 0, \quad 0 < x < L,$$

$$u(x, 0) = \sin(x), \quad 0 < x < L.$$

Assume that we know an appropriate boundary condition is $t\cos t$, where should it be prescribed, $x = 0$ or $x = L$? Solve the problem as well.

Solution: In this example $a = -2$, so the slope of characteristics is negative. The line passing through $x = 0$ and $t = 0$ is $x + 2t = 0$ that is out of the solution domain. The line passing through $x = L$ and $t = 0$ with slope -2 is $x + 2t = C$. Plugging in $x = L$ and $t = 0$, we get $C = L$. The line $x + 2t = L$ divides the solution domain in two regions. The solution in the region bounded by $x = 0$, $t = 0$, and $x + 2t = L$, $0 < x < L$, can trace back to the initial condition.

In another region, the solution can be traced back to the boundary condition at $x = L$. The line equation that passes through (x, t) with slope -2 can be written as $(X - x) + 2(T - t) = 0$. Let the intersection of the line with $x = L$ be (L, t^*).

Plug them into the line equation to get $(L-x) + 2(t^* - t) = 0$ and solve for t^* to get $t^* = t - (L-x)/2$. Thus, the boundary condition should be described at $x = L$ as $u(L, t) = t \cos t$. The solution is

$$u_0(x,t) = \begin{cases} \sin(x+2t) & 0 < x < L+2t;\ t > 0, \\ \left(t - \dfrac{L-x}{2}\right) \cos\left(t - \dfrac{L-x}{2}\right) & \max\{0, L-2t\} < x < L,\ t > 0. \end{cases}$$

2.6 Method of characteristics for general linear first order PDEs

Consider a general linear and homogeneous first order PDE

$$\frac{\partial u}{\partial t} + p(x,t) \frac{\partial u}{\partial x} = 0. \tag{2.22}$$

Using the method of characteristics, we set $\frac{dx}{dt} = p(x,t)$. If we can solve this ODE to get $x - g(t) = C$. Then the general solution to the original problem is $u(x,t) = f(x - g(t))$ for any differentiable function $f(x)$.

Proof: If $u(x,t) = f(x - g(t))$ and $\frac{dx}{dt} = -g'(t) = p(x,t)$, then we have $\frac{\partial u}{\partial t} = f' g'(t) = -f' p(x,t)$ and $\frac{\partial u}{\partial x} = f'$. Thus we have $\frac{\partial u}{\partial t} + p(x,t) \frac{\partial u}{\partial x} = f'(-p) + pf' = 0$.

General Solution to an Advection Equation with a Variable Coefficient $\frac{\partial u}{\partial t} + p(x,t)\frac{\partial u}{\partial x} = 0$. Use one of two below.

$$\frac{dx}{dt} = p(x,t),\ x = \int p\, dt + C = f(x,t) + C \Longrightarrow u(x,t) = G(x - f(x,t)).$$

or $\dfrac{dt}{dx} = \dfrac{1}{p(x,t)},\ t = \int \dfrac{1}{p} dx + C = r(x,t) + C \Longrightarrow u(x,t) = G(t - r(x,t)).$

Example 2.5. *Find the general solution to*

$$\frac{\partial u}{\partial t} + x^2 \frac{\partial u}{\partial x} = 0.$$

Find also the solution to the Cauchy problem if $u(x, 0) = \sin x$.

Solution: We set $\frac{dx}{dt} = p(x,t) = x^2$ or $\frac{dx}{x^2} = dt$. We get $-\frac{1}{x} = t + C$ or $C = t + \frac{1}{x}$. The general solution is $u(x,t) = f(t + \frac{1}{x})$.

Since we have $u(x,0) = f(-1/x) = \sin x$. Let $y = 1/x$ we get $f(y) = \sin y$. The solution to the Cauchy problem is $u(x,t) = \sin \frac{1}{t+1/x} = \sin \frac{x}{tx+1}$. We verified that the solution satisfies the PDE using the Maple.

Example 2.6. *Find the general solution to*
$$\frac{1}{t^2}\frac{\partial u}{\partial t} + \frac{\partial u}{\partial x} = 0.$$
Find also the solution to the Cauchy problem if $u(x,0) = \sin x$.

Solution: We set $\frac{dx}{dt} = p(x,t) = t^2$ or $x = t^3/3 + C$. We get $C = x - t^3/3$. The general solution is $u(x,t) = f(x - \frac{t^3}{3})$.

Since we have $u(x,0) = f(x) = \sin x$. The solution to the Cauchy problem is $u(x,t) = \sin\left(x - \frac{t^3}{3}\right)$, which satisfies the PDE as verified by the Maple.

2.7 Solution to first order linear non-homogeneous PDEs with constant coefficients

Using the method of changing variables, we can transform a first order linear non-homogeneous PDEs with constant coefficients

$$\frac{\partial u}{\partial t} + a\frac{\partial u}{\partial x} + bu = f(x,t) \qquad (2.23)$$

to an ODE. Thus we can solve the ODE to get the general solution to the PDE. We use the same new variables

$$\begin{cases} \xi = x - at, \\ \eta = t, \end{cases} \quad \text{or} \quad \begin{cases} x = \xi + a\eta, \\ t = \eta. \end{cases} \qquad (2.24)$$

Under such a transform, we have $u(x,t) = u(\xi + a\eta, \eta)$. We denote $U(\xi, \eta) = u(\xi + a\eta, \eta)$. Then using the chain rule, we can get

$$\frac{\partial u}{\partial t} = \frac{\partial U}{\partial \xi}\frac{\partial \xi}{\partial t} + \frac{\partial U}{\partial \eta}\frac{\partial \eta}{\partial t} = -a\frac{\partial U}{\partial \xi} + \frac{\partial U}{\partial \eta},$$

$$\frac{\partial u}{\partial x} = \frac{\partial U}{\partial \xi}\frac{\partial \xi}{\partial x} + \frac{\partial U}{\partial \eta}\frac{\partial \eta}{\partial x} = \frac{\partial U}{\partial \xi}.$$

Plug them into the original PDE (2.23), we would get

$$\frac{\partial U}{\partial \eta} + bU = f(\xi + a\eta, \eta) = F(\xi, \eta). \qquad (2.25)$$

2.7. Solution to first order linear non-homogeneous PDEs

The equation above is actually ordinary differential equation with respect to η (treating ξ as a constant). If we can solve the ODE above, we can get the general solution to the original PDE.

Example 2.7. *Find the general solution to*

$$\frac{\partial u}{\partial t} + 2\frac{\partial u}{\partial x} - u = t.$$

Solution: With the changing variable $\xi = x - 2t$, $\eta = t$, the PDE becomes

$$\frac{\partial U}{\partial \eta} - U = \eta.$$

It is a non-homogeneous ODE and the solution can be expressed as

$$U = U_h + U_p$$

in which U_h is the homogeneous solution to $\frac{\partial U}{\partial \eta} - U = 0$ and U_p is a particularly solution to the ODE. It is easy to get $U_h(\xi, \eta) = g(\xi)e^\eta$. From the ODE technique, we can set

$$U_p = A\eta + B$$

for two constants A and B. Plug this into the ODE and matching terms on both sides, we get $A = -1$, $B = -1$. Thus the solution in the new variables is

$$U_h(\xi, \eta) = g(\xi)e^\eta - \eta - 1.$$

Thus, the general solution to the PDE then is

$$u(x, t) = g(x - 2t)e^t - t - 1.$$

Solution to First Order non-Homogeneous PDE with Constant Coefficients $\frac{\partial u}{\partial t} + a\frac{\partial u}{\partial x} + bu = f(x, t)$.

$\xi = x - at$, $\eta = t$, $\implies$ $\frac{\partial U}{\partial \eta} + bU = F(\xi, \eta)$, Assume the solution is

$U(\xi, \eta)$, then the original solution is $u(x, t) = U(x - at, t)$.

Example 2.8. *Find the general solution to*

$$\frac{\partial u}{\partial t} - \frac{1}{2}\frac{\partial u}{\partial x} + 4u = xt.$$

Solution: With the changing variable $\xi = x + \frac{1}{2}t$, $\eta = t$, the PDE becomes

$$\frac{\partial U}{\partial \eta} + 4U = \left(\xi - \frac{1}{2}\eta\right)\eta.$$

It is a non-homogeneous ODE and the solution can be expressed as

$$U = U_h + U_p$$

in which U_h is the homogeneous solution to $\frac{\partial U}{\partial \eta} + 4U = 0$ and U_p is a particularly solution to the ODE. It is easy to get $U_h(\xi, \eta) = g(\xi)e^{-4\eta}$. From the ODE technique, we can set

$$U_p = A\eta^2 + B\eta + C$$

where A, B, and C are constants. Plug this into the ODE and matching terms on both sides, we get $A = -1/8$, $B = \frac{\xi}{4} + \frac{1}{16}$, $C = -\frac{\xi}{16} - \frac{1}{64}$. Thus the solution in the new variables is

$$U_h(\xi, \eta) = g(\xi)e^{-4\eta} - \frac{\eta^2}{8} + \left(\frac{\xi}{4} + \frac{1}{16}\right)\eta - \frac{\xi}{16} - \frac{1}{64}.$$

Thus the general solution to the PDE then is

$$u(x,t) = g\left(x + \frac{t}{2}\right)e^{-4t} - \frac{t^2}{8} + \left(\frac{x+t/2}{4} + \frac{1}{16}\right)t - \frac{x+t/2}{16} - \frac{1}{64}.$$

2.8 Exercises

E2.1 Classify the following PDE as much as you can (linear, quasi-linear, or non-linear; order; constant or variable coefficient(s); homogeneous or not; dimension(s); type: hyperbolic, elliptic, parabolic; physical meanings: heat, wave, potential) as much as you can. Also give physical backgrounds if you can.

(a)
$$Au_t = B(u_{xx} + u_{yy}) + f(x,y), \quad \text{consider } A \neq 0 \text{ and } A = 0.$$

(b)
$$u_{tt} = B(u_{xx} + u_{yy}) + f(x,y)$$

2.8. Exercises

(c)
$$u_t + au_x = Bu_{xx} + f(u).$$

In the expressions above, A, B, a, and μ are constants.

E2.2 Given $\dfrac{\partial u}{\partial t} + \dfrac{\partial u}{\partial x} = 0.$

(a) Find the general solution.

(b) Find the solution to the Cauchy problem given $u(x,0) = 2e^{-2x^2}$, $-\infty < x < \infty$, $t > 0$.

(c) Sketch of the solution at $t = 3$ if the initial condition $u(x,0)$ is given

$$u(x,0) = \begin{cases} 1 - |x| & -1 \leq x \leq 1, \\ 0 & \text{otherwise}, \end{cases}$$

see the top plot which is called a hat function. Mark the height of the solution at $t = 3$.

E2.3 Derive the general solution of the given equation

(a), $2u_t + 3u_x = 0;$ (b), $au_t + bu_x = u$, $a^2 + b^2 \neq 0.$

Solve the Cauchy problem with $u(x,0) = \sin x$, and $u(x,0) = e^{-x^2}$.

E2.4 Solve the given partial differential equations below by the method of characteristics. Check you your answer by plugging it back into the equation.

(a)
$$u_t + \sin t\, u_x = 0.$$

(b)
$$e^{x^2} u_x + x u_y = 0.$$

E2.5 Find the solution to the following transport equation:

$$\dfrac{\partial u}{\partial t} + \dfrac{1}{2}\dfrac{\partial u}{\partial x} = 0, \quad 0 < x < 1, \quad t > 0,$$
$$u(x,0) = e^{-x}, \quad 0 < x < 1,$$
$$u(0,t) = t^2, \quad 0 < t.$$

E2.6 Find the solution to the following transport equation * :

$$\frac{\partial u}{\partial t} - \frac{1}{2}\frac{\partial u}{\partial x} = 0, \quad 0 < x < 1, \quad t > 0,$$
$$u(x,0) = e^{-x}, \quad 0 < x < 1,$$
$$u(1,t) = t^2, \quad 0 < t.$$

E2.7 Solve the following PDE with $u(x,0) = f(x)$.

(a) $u_t + au_x = e^{2x}$.

(b) $u_t + xu_x = 0$.

(c) $u_t + tu_x = 0$.

(d) $u_t + 3u_x = u + xt$.

E2.8 Simulate the solution (make plots or movies) using Maple or Matlab for the advection equation $u_t + au_x = 0$ of the Cauchy problem ($-\infty < x < \infty$) with the following initial conditions:

(a)
$$u_0(x) = \begin{cases} \cos(2x) & -2\pi \leq x \leq 2\pi, \\ 0 & \text{otherwise.} \end{cases}$$

(b)
$$u_0(x) = \begin{cases} 1 - |x| & -1 \leq x \leq 1 \\ 0 & \text{otherwise.} \end{cases}$$

Chapter 3

Solution to one-dimensional wave equations

A one-dimensional (1D) wave equation has the following form

$$\frac{\partial^2 u}{\partial t^2} = c^2 \frac{\partial^2 u}{\partial x^2} \tag{3.1}$$

where $c > 0$ is the wave speed in physics. The partial differential equations is a second order, linear, constant coefficient, homogeneous one. According to the criterion defined on page 6, the PDE is classified as hyperbolic. We first to derive the general solution for which no constraints are imposed.

We can use the method of changing variables to simplify the PDE by setting

$$\begin{cases} \xi = x - ct, \\ \eta = x + ct, \end{cases} \quad \text{or} \quad \begin{cases} x = \dfrac{\xi + \eta}{2}, \\ t = \dfrac{\eta - \xi}{2c}. \end{cases} \tag{3.2}$$

Under such a transform, we have $u(x,t) = u\left(\frac{\xi+\eta}{2}, \frac{\eta-\xi}{2c}\right) = U(\xi, \eta)$. Then using the chain rule, we can get

$$\frac{\partial u}{\partial t} = \frac{\partial U}{\partial \xi}\frac{\partial \xi}{\partial t} + \frac{\partial U}{\partial \eta}\frac{\partial \eta}{\partial t} = -c\frac{\partial U}{\partial \xi} + c\frac{\partial U}{\partial \eta},$$

$$\frac{\partial u}{\partial x} = \frac{\partial U}{\partial \xi}\frac{\partial \xi}{\partial x} + \frac{\partial U}{\partial \eta}\frac{\partial \eta}{\partial x} = \frac{\partial U}{\partial \xi} + \frac{\partial U}{\partial \eta}.$$

Differentiating the first expressions above with respect to t again, we get

$$\frac{\partial^2 u}{\partial t^2} = (-c)\frac{\partial^2 U}{\partial \xi^2}\frac{\partial \xi}{\partial t} + (-c)\frac{\partial^2 U}{\partial \eta \partial \xi}\frac{\partial \eta}{\partial t} + c\frac{\partial^2 U}{\partial \xi \partial \eta}\frac{\partial \xi}{\partial t} + c\frac{\partial^2 U}{\partial \eta^2}\frac{\partial \eta}{\partial t}$$

$$= c^2 \frac{\partial^2 U}{\partial \xi^2} - 2c^2 \frac{\partial^2 U}{\partial \xi \partial \eta} + c^2 \frac{\partial^2 U}{\partial \eta^2}, \tag{3.3}$$

assuming both $\frac{\partial^2 U}{\partial \xi \partial \eta}$ and $\frac{\partial^2 U}{\partial \eta \partial \xi}$ are continuous so that they are the same. Similarly we can derive

$$\frac{\partial^2 u}{\partial x^2} = \frac{\partial^2 U}{\partial \xi^2} + 2\frac{\partial^2 U}{\partial \xi \partial \eta} + \frac{\partial^2 U}{\partial \eta^2}. \tag{3.4}$$

Plugging (3.3) and (3.4) into the original PDE, we obtain

$$4c^2 \frac{\partial^2 U}{\partial \xi \partial \eta} = 0 \quad \text{or} \quad \frac{\partial^2 U}{\partial \xi \partial \eta} = 0$$

after some manipulations and using the fact that $c \neq 0$. We integrate $\frac{\partial^2 U}{\partial \xi \partial \eta} = 0$ with respect to η to get $\frac{\partial U}{\partial \xi} = f(\xi)$; and integrate it with respect to ξ to further have

$$U(\xi, \eta) = \int f(\xi) d\xi + G(\eta) = F(\xi) + G(\eta)$$

since $\int f(\xi) d\xi$ is still a function of ξ. Finally, we substitute the new variables back to the original ones to get the general solution

$$u(x,t) = u\left(\frac{\xi + \eta}{2}, \frac{\eta - \xi}{2c}\right) = U(\xi, \eta) = F(x - ct) + G(x + ct) \tag{3.5}$$

for any twice one-dimensional differentiable functions $F(x)$ and $G(x)$.

The General Solution of 1D Wave Equation $\frac{\partial^2 u}{\partial t^2} = c^2 \frac{\partial^2 u}{\partial x^2}$ **is**

$$u(x,t) = F(x - ct) + G(x + ct)$$

for any differentiable function $F(x)$ **and** $G(x)$.

Example 3.1. *The general solution to* $2\frac{\partial^2 u}{\partial t^2} - 3\frac{\partial^2 u}{\partial x^2} = 0$ *is*

$$u(x,t) = F\left(x + \sqrt{\frac{3}{2}}t\right) + G\left(x - \sqrt{\frac{3}{2}}t\right)$$

for any differentiable function $F(x)$ *and* $G(x)$.

3.1 Solution to Cauchy problems of 1D wave equations

A Cauchy problem (an initial value problem) of a one-dimensional (1D) wave equation has the following form

$$\frac{\partial^2 u}{\partial t^2} = c^2 \frac{\partial^2 u}{\partial x^2}, \quad -\infty < x < \infty, \tag{3.6}$$

$$u(x,0) = f(x), \quad \frac{\partial u}{\partial t}(x,0) = g(x), \quad -\infty < x < \infty, \tag{3.7}$$

where $f(x)$ and $g(x)$ are given initial conditions. The solution to the Cauchy problem can be represented by the D'Alembert's formula

$$u(x,t) = \frac{1}{2}\left(f(x-at) + f(x+ct)\right) + \frac{1}{2c}\int_{x-at}^{x+ct} g(s)ds. \tag{3.8}$$

Proof: First we check the initial conditions. We have

$$u(x,0) = \frac{1}{2}\left(f(x) + f(x)\right) + 0 = f(x)$$

since the integration is zero if the lower and upper limits of the integration are the same. Secondly, we differentiate the equality (3.8) with respect to t first, then let $t = 0$ to get

$$\frac{\partial u}{\partial t}(x,0) = \frac{1}{2}\left(f'(x)(-c) + f'(x)c\right) + \frac{1}{2c}\left(cg(x) - g(x)(-c)\right) = g(x).$$

To prove that the function in the D'Alembert's formula satisfies the wave equation, we just need to find $F(x)$ and $G(x)$ in the general solution in terms of $f(x)$ and $g(x)$. From the initial condition, we already have

$$u(x,0) = F(x) + G(x) = f(x). \tag{3.9}$$

Differentiating the general solution $u(x,t) = F(x-ct) + G(x+ct)$ with respect to t, we get

$$\frac{\partial u}{\partial t} = F'(x-ct)(-c) + G'(x+ct)c, \tag{3.10}$$

which leads to

$$\frac{\partial u}{\partial t}(x,0) = -F'(x)c + G'(x)c = g(x). \tag{3.11}$$

We further integrate the equality above from 0 to x to get

$$-F(x) + G(x) = \frac{1}{c}\int_0^x g(s)ds + 2A, \tag{3.12}$$

where A is a constant. Note that we use $2A$ for simplicity of derivation. Add (3.12) and $F(x) + G(x) = f(x)$ together we get

$$G(x) = \frac{1}{2}f(x) + \frac{1}{2c}\int_0^x g(s)ds + A.$$

From (3.9) and the above identity, we also arrive at

$$F(x) = f(x) - G(x) = \frac{1}{2}f(x) - \frac{1}{2c}\int_0^x g(s)ds - A.$$

Plug $F(x)$ and $G(x)$ above into the general solution, we get

$$u(x,t) = F(x-ct) + G(x+ct) = \frac{1}{2}f(x-ct) - \frac{1}{2c}\int_0^{x-ct} g(s)ds - A$$

$$+ \frac{1}{2}f(x+ct) + \frac{1}{2c}\int_0^{x+ct} g(s)ds + A$$

$$= \frac{1}{2}\left(f(x+ct) + f(x+ct)\right) + \frac{1}{2c}\int_{x-ct}^0 g(s)ds + \frac{1}{2c}\int_0^{x+ct} g(s)ds$$

$$= \frac{1}{2}\left(f(x-ct) + f(x+ct)\right) + \frac{1}{2c}\int_{x-ct}^{x+ct} g(s)ds.$$

Solution to the Cauchy problem of a 1D Wave Equation

$$\frac{\partial^2 u}{\partial t^2} = c^2 \frac{\partial^2 u}{\partial x^2}, \qquad -\infty < x < \infty,$$

$$u(x,0) = f(x), \quad \frac{\partial u}{\partial t}(x,0) = g(x), \qquad -\infty < x < \infty,$$

is given by the D'Alembert's formula

$$u(x,t) = \frac{1}{2}\left(f(x-at) + f(x+ct)\right) + \frac{1}{2c}\int_{x-at}^{x+ct} g(s)ds.$$

Example 3.2. *Solve the Cauchy problem for the wave equation*

$$\frac{\partial^2 u}{\partial t^2} = 4\frac{\partial^2 u}{\partial x^2}, \qquad -\infty < x < \infty,$$

$$u(x,0) = \begin{cases} \sin x & \text{if } |x| \le \pi, \\ 0 & \text{otherwise,} \end{cases} \qquad \frac{\partial u}{\partial t}(x,0) = 0.$$

3.1. Solution to Cauchy problems of 1D wave equations

Solution: The solution is simply

$$u(x,t) = \frac{1}{2}\left(f(x-2t) + f(x+2t)\right).$$

If t is large enough, then the non-zero regions of $f(x-2t)$ and $f(x+2t)$ do not overlap. We see clearly a single sine wave in the domain $(-\pi, \pi)$ propagates to the right and left with half the magnitude, see the solution plot in Fig. 3.1 for an illustration. In the literature $f(x-ct)$ is called the right-going wave, while $f(x+ct)$ the left-going wave.

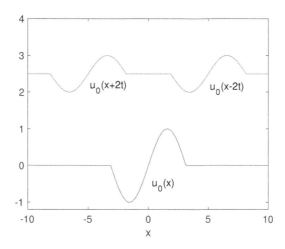

Figure 3.1. *Plot of the initial condition $u_0(x)$ and a solution $u(x,t)$ at a later time ($t = 2.5$) to the 1D wave solution with the wave speed $c = 2$.*

Example 3.3. *Solve the Cauchy problem for the wave equation*

$$\frac{\partial^2 u}{\partial t^2} = 2\frac{\partial^2 u}{\partial x^2}, \quad -\infty < x < \infty$$

$$u(x,0) = \sin x \quad \frac{\partial u}{\partial t}(x,0) = xe^{-x^2}.$$

Solution: According to the D'Alembert's formula, the solution is

$$u(x,t) = \frac{1}{2}\left(\sin(x-\sqrt{2}t) + \sin(x+\sqrt{2}t)\right) + \frac{1}{2\sqrt{2}}\int_{x-\sqrt{2}t}^{x+\sqrt{2}t} se^{-s^2}\,ds$$

$$= \frac{1}{2}\left(\sin(x-\sqrt{2}t) + \sin(x+\sqrt{2}t)\right) + \frac{1}{4\sqrt{2}}\left(e^{-(x-\sqrt{2}t)^2} - e^{-(x+\sqrt{2}t)^2}\right).$$

Example 3.4. *Plot or sketch of the solution of the Cauchy problem for the wave equation below for large t.*

$$\frac{\partial^2 u}{\partial t^2} = \frac{\partial^2 u}{\partial x^2}, \quad -\infty < x < \infty,$$

$$u(x,0) = \begin{cases} 2x & \text{if } 0 \leq x \leq \frac{1}{2}, \\ 2(1-x) & \text{if } \frac{1}{2} \leq x \leq 1, \\ 0 & \text{otherwise.} \end{cases} \quad \frac{\partial u}{\partial t}(x,0) = 0.$$

In Figure 3.2, we show the plot of the solution at time $t = 0$, $t = 0.3$, and $t = 5$. We can see clearly how the one wave split into two with half strength towards left $(x - t))$ and right $(x - t))$. A Matlab movie file is also available (wave_piece.m and fp.m). For this kind of problems when the initial condition is a piecewise continuous function, it is much easy to use a computer to find and plot the solution.

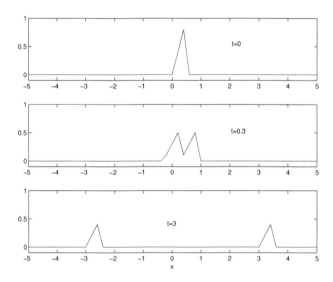

Figure 3.2. *Plot of the wave propagation at time $t = 0$, $t = 0.3$, and $t = 5$. We can see clearly how the one wave split into two with half strength towards left $(x - t)$ and right $(x - t)$.*

3.2 Normal modes solutions to 1D wave equations of BVPs with special initial conditions

Now consider the boundary value problem of an one dimensional wave equation

$$\frac{\partial^2 u}{\partial t^2} = c^2 \frac{\partial^2 u}{\partial x^2}, \quad 0 < x < L$$
$$u(0,t) = 0, \quad u(L,t) = 0, \tag{3.13}$$
$$u(0,t) = f(x), \quad \frac{\partial u}{\partial t}(x,0) = g(x), \quad 0 < x < L,$$

for a positive constant L. An application is an elastic string of a length L with two ends fixed, which corresponds to the homogeneous boundary condition, see Fig. 3.3 for an illustration.

Figure 3.3. *A diagram of an elastic string with two ends fixed. The motion can be modeled using a 1D wave equation.*

Since the motion of an elastic string is oscillatory, we would expect the solution is of sort of trigonometric functions of two variables of x and t. We can consider $\sin(\alpha x)\cos(\beta t)$, $\sin(\alpha x)\sin(\beta t)$, $\cos(\alpha x)\cos(\beta t)$, $\cos(\alpha x)\sin(\beta t)$, etc. The vanishing boundary condition at $x = 0$ eliminates the $\cos(\beta x)$ option. The solution would look like $\sin(\alpha x)\cos(\beta t)$, $\sin(\alpha x)\sin(\beta t)$. Since the solution is zero at $x = L$, we should have $\sin(\alpha L) = 0$ which means $\alpha L = n\pi$ for $n = 1, 2, \cdots$. Finally the solution should satisfy the PDE, after we differentiate $\sin(\alpha x)\cos(\beta t)$ twice with x and t, we will get $\beta = nc\pi/L$.

Thus, a special function

$$u_n(x,t) = \sin\frac{n\pi x}{L}\cos\frac{n\pi ct}{L}, \tag{3.14}$$

for a non-zero integer n can be one of solutions. It is obviously that $u_n(0,t) = u(L,t) = 0$ and $u_n(x,t)$ satisfies the PDE (3.13). Note that $u_n(x,0) = \sin\frac{n\pi x}{L}$ and $\frac{\partial u}{\partial t}(x,0) = 0$. Thus, if $f(x) = \sin\frac{n\pi x}{L}$ and $g(x) = 0$, then $u_n(x,t)$ is the solution to the initial-boundary value problem (3.13). Such a solution is called a normal mode solution to the initial-boundary value problem.

Similarly,
$$\bar{u}_n(x,t) = \sin\frac{n\pi x}{L}\sin\frac{n\pi ct}{L} \qquad (3.15)$$
also satisfies the boundary condition and the PDE. Now we have $\bar{u}_n(x,0) = 0$ and $\frac{\partial \bar{u}}{\partial t}(x,0) = \frac{nc\pi}{L}\sin\frac{n\pi x}{L}$. Thus, if $f(x) = 0$ and $g(x) = \frac{nc\pi}{L}\sin\frac{n\pi x}{L}$, then $\bar{u}_n(x,t)$ is a normal mode solution to the initial-boundary value problem (3.13).

The following normal modes

$$\sin\frac{n\pi x}{L}\sin\frac{n\pi ct}{L}, \quad \sin\frac{n\pi x}{L}\cos\frac{n\pi ct}{L},$$

$$\sum_{n=1}^{N} \sin\frac{n\pi x}{L}\left(A_n \sin\frac{n\pi ct}{L} + B_n \cos\frac{n\pi ct}{L}\right),$$

satisfy the 1D wave equation and the homogeneous boundary conditions, but NOT arbitrary initial conditions.

Example 3.5. If $f(x) = \frac{1}{2}\sin\frac{5\pi x}{L}$ and $g(x) = 0$, find the solution to the initial-boundary value problem (3.13).

Solution: According to the normal mode solution, the solution is
$$u(x,t) = \frac{1}{2}\sin\frac{5\pi x}{L}\cos\frac{5\pi ct}{L}.$$

Example 3.6. If $f(x) = 0$ and $g(x) = \frac{1}{2}\sin\frac{5\pi x}{L}$, find the solution to the initial-boundary value problem (3.13).

Solution: According to the normal mode solution of the second type, the solution is
$$u(x,t) = \frac{L}{10\pi c}\sin\frac{5\pi x}{L}\sin\frac{5\pi ct}{L}.$$

Example 3.7. If $f(x) = \sin\frac{5\pi x}{L} - 10\sin\frac{20\pi x}{L}$ and $g(x) = \frac{1}{2}\sin\frac{15\pi x}{L}$, find the solution to the initial-boundary value problem (3.13).

Solution: *According to the normal mode solutions and the principle of superposition, the solution is*

$$u(x,t) = \sin\frac{5\pi x}{L}\cos\frac{5\pi ct}{L} - 10\sin\frac{20\pi x}{L}\cos\frac{20\pi ct}{L} + \frac{L}{30\pi c}\sin\frac{15\pi x}{L}\sin\frac{15\pi ct}{L}.$$

This is because the PDE is linear, homogeneous, and with homogeneous boundary conditions.

Challenge: How about the normal modes of

$$\tilde{u}_n(x,t) = \cos\frac{n\pi x}{L}\sin\frac{n\pi ct}{L}, \qquad \hat{u}_n(x,t) = \cos\frac{n\pi x}{L}\cos\frac{n\pi ct}{L}?$$

From the superposition, we know that the linear combination

$$u_N(x,t) = \sum_{n=1}^{N}\left(a_n \sin\frac{n\pi x}{L}\cos\frac{n\pi ct}{L} + b_n \sin\frac{n\pi x}{L}\sin\frac{n\pi ct}{L}\right) \qquad (3.16)$$

is the solution to the initial-boundary value problem (3.13) with special initial condition

$$u_N(x,0) = f(x) = \sum_{n=1}^{N} a_n \sin\frac{n\pi x}{L},$$

$$\frac{\partial u_N}{\partial t}(x,0) = g(x) = \sum_{n=1}^{N} b_n \frac{L}{n\pi c}\sin\frac{n\pi x}{L}.$$

What should we do for other general $f(x)$ and $g(x)$? We can use the method of separation variablea and Fourier expansions ($N \to \infty$) that will be discussed later.

3.3 Exercises

E3.1 Let $\dfrac{\partial^2 u}{\partial t^2} = c^2 \dfrac{\partial^2 u}{\partial x^2}$. **(A)** Find the general solution; **(B)** Find the solution to the Cauchy problem given $u(x,0) = f(x)$ and $\dfrac{\partial u}{\partial t}(x,0) = g(x)$.

(a) $c = 1/\pi$, $f(x) = \sin(\pi x)$, $g(x) = 0$.
(b) $c = 1$, $f(x) = \sin(\pi x) + 3\sin(2\pi x)$, $g(x) = \sin(\pi x)$.

(c) Computer project. (**Extra Credit**) $c = 1$, $g(x) = x$,

$$f(x) = \begin{cases} 2x & \text{if } 0 \leq x \leq \frac{1}{2}, \\ 2(1-x) & \text{if } \frac{1}{2} \leq x \leq 1, \\ 0 & \text{otherwise.} \end{cases}$$

Plot or sketch the solution at $t = 0.5$ and 1 for all the problems above. Make a movie of the solution between $0 \leq t \leq 50$.

E3.2 Let $\dfrac{\partial^2 u}{\partial t^2} = c^2 \dfrac{\partial^2 u}{\partial x^2}$, $0 < x < L$. with $u(0,t) = u(L,t) = 0$. Find the solution to the initial and boundary value problem given $u(x,0) = f(x)$ and $\dfrac{\partial u}{\partial t}(x,0) = g(x)$.

(a) $f(x) = \sin \dfrac{2\pi x}{L}$, $g(x) = 0$.

(b) $f(x) = \dfrac{1}{2} \sin \dfrac{\pi x}{L} + \dfrac{1}{4} \sin \dfrac{3\pi x}{L} + \dfrac{2}{5} \sin \dfrac{7\pi x}{L}$, $g(x) = 0$.

(c) $f(x) = 0$, $g(x) = \dfrac{1}{4} \sin \dfrac{3\pi x}{L} - \dfrac{1}{10} \sin \dfrac{6\pi x}{L}$.

(d) $f(x) = \dfrac{1}{4} \sin \dfrac{3\pi x}{L} + \dfrac{1}{10} \sin \dfrac{6\pi x}{L}$, $g(x) = \dfrac{1}{4} \sin \dfrac{3\pi x}{L} - \dfrac{2}{5} \sin \dfrac{7\pi x}{L}$.

E3.3 Assume that $c = 2$, $L = 3$, can you solve the 1D wave equation (3.13) with the following $f(x)$ and $g(x)$?

(a) $f(x) = \sin(6\pi x)$, $g(x) = 0$.

(b) $f(x) = 0$, $g(x) = \sin(3\pi x)$.

(c) $f(x) = x \sin x$, $g(x) = 0$.

(d) $f(x) = \sin(6\pi x)$, $g(x) = \sin(3\pi x)$.

(e) $f(x) = \sin(6\pi x) - 7\sin(24\pi x)$, $g(x) = \sin(3\pi x)$.

E3.4 Given the functions below

$$u_1(x,t) = \sum_{n=1}^{N} \dfrac{1}{n} \sin \dfrac{n\pi x}{2} \sin \dfrac{n\pi \sqrt{3} t}{2}, \qquad u_2(x,t) = \sum_{n=1}^{N} \dfrac{1}{n} \sin \dfrac{n\pi x}{2} \cos \dfrac{n\pi \sqrt{3} t}{2}.$$

(a) Assume that $u(x,0) = u_1(x,0)$ and $\dfrac{\partial u}{\partial t}(x,0) = u_2(x,0)$, and $u(0,t) = u(2,t) = 0$, find the wave equation and the solution with those initial and boundary conditions.

(b) Use a computer software (Maple, Matlab, Mathematica etc.) to plot the functions with $n = 5$, $n = 50$, and $n = 500$. What do you observe?

Chapter 4

Orthogonal functions & expansions, and Sturm-Liouville theory

For one-dimensional wave equations with homogeneous boundary conditions, even if there is only one non-zero initial condition that is an elementary function such as $u(x,0) = f(x) = e^x$ or $f(x) = \sum_{i=0}^{N} a_i x^i$ assuming that $u_t(x,0) = 0$, we cannot use a combination of normal mode solutions unless we have infinite number of terms. In fact, we may be able to get a series solution if we can expand e^x, for example, as below

$$e^x \sim \sum_{n=1}^{\infty} b_n \sin \frac{n\pi x}{L}. \tag{4.1}$$

We use the symbol $\sim$ to indicate that the above expansion may or may not be identical on both sides. Is this expansion possible? When is this possible? Is this valid in $(0, L)$ or any interval? If the expansion is valid in $(0, L)$, then (4.1) is called an orthogonal functions expansion of e^x in terms of $\left\{ \sin \frac{n\pi x}{L} \right\}$. Often we use the pair of braces '{ }' to represent a set. How do we get those orthogonal functions? One of answers is from the Sturm-Liouville eigenvalue theory.

Here we give a glimpse of the method of separation of variables for a 1D wave equation of a boundary value problem,

$$\frac{\partial^2 u}{\partial t^2} = c^2 \frac{\partial^2 u}{\partial x^2}, \quad 0 < x < L,$$

$$u(0,t) = 0, \quad u(L,t) = 0,$$

$$u(0,t) = f(x), \quad \frac{\partial u}{\partial t}(x,0) = g(x), \quad 0 < x < L.$$

We try a solution that has the special form $u(x,t) = T(t)X(x)$, in which the variables are separated. To satisfy the boundary conditions, we should have $X(0) = 0$

and $X(L) = 0$ since $T(t)$ cannot be zero. Thus, we have $\frac{\partial^2 u}{\partial t^2} = T''(t)X(x)$ and $\frac{\partial^2 u}{\partial x^2} = T(t)X''(x)$. Plugging them into the wave equation we get

$$T''(t)X(x) = c^2 T(t)X''(x) \implies \frac{T''(t)}{c^2 T(t)} = \frac{X''(x)}{X(x)}.$$

In the second expression of above, the left hand side is a function of t while the right hand side is a function of x, which is possible only if

$$\frac{T''(t)}{c^2 T(t)} = \frac{X''(x)}{X(x)} = -\lambda$$

for some constant λ. We will see why we use the negative sign in front of λ later. Thus, we have

$$X''(x) + \lambda X(x) = 0, \qquad X(0) = X(L) = 0. \tag{4.2}$$

This is called a Sturm-Liouville eigenvalue problem of the boundary value problem since both λ and $u(x)$ are unknowns. From ordinary differential equation solution methods and theory we know that the solution is

$$X(x) = C_1 e^{\sqrt{-\lambda}\,x} + C_2 e^{-\sqrt{-\lambda}\,x}.$$

If $\lambda \leq 0$, then we have to have $X(x) = 0$ from the boundary condition. $X(x) = 0$ is called a *trivial solution*. Note also that $X(x) = 0$ cannot satisfy the initial condition unless $f(x) = 0$ and $g(x) = 0$, and thus, should be discarded. If $\lambda > 0$, then the solution is

$$X(x) = C_1 \cos(\sqrt{\lambda}\,x) + C_2 \sin(\sqrt{\lambda}\,x).$$

The condition $X(0) = 0$ implies that $C_1 = 0$. Thus, we get $X(x) = C_2 \sin(\sqrt{\lambda}x)$. The condition $X(L) = 0$ implies that $X(L) = \sin(\sqrt{\lambda}L) = 0$, which leads to

$$\sqrt{\lambda}\,L = n\pi, \quad \text{or} \quad \lambda = \left(\frac{n\pi}{L}\right)^2, \quad n = 1, 2, \cdots,$$

$$\text{and} \quad X_n(x) = \sin\frac{n\pi x}{L}.$$

The functions $\{X_n(x)\} = \{\sin\frac{n\pi x}{L}\}$ above satisfy the ODE and the homogeneous boundary conditions for any natural number n, and are called the eigenfunctions of the special Sturm-Liouville eigenvalue problem. Note that, we usually ignore the constant C_2 in the expression of $\{X_n(x)\}$ because eigenfunctions can differ by a constant. More important, those eigenfunctions are normal modes as we know. Next, we solve for $T(t)$ using

$$T''(t) + c^2 \lambda_n T(t) = 0, \tag{4.3}$$

where $\lambda_n = (\frac{n\pi}{L})^2$ that have already been found. Therefore, the solution of $T(t)$ is,

$$T_n(t) = b_n \cos(\sqrt{\lambda}\, ct) + b_n^* \sin(\sqrt{\lambda}\, ct) = b_n \cos \frac{n\pi ct}{L} + b_n^* \sin \frac{n\pi ct}{L}.$$

We put $X_n(x)$ and $T_n(t)$ together to get a normal mode solution

$$u_n(x,t) = \sin \frac{n\pi x}{L} \left(b_n \cos \frac{cn\pi t}{L} + b_n^* \sin \frac{cn\pi t}{L} \right), \tag{4.4}$$

which satisfy the PDE, the boundary conditions, but not the initial conditions.

We put all the normal mode solutions together to get a series solution.

$$u(x,t) = \sum_{n=0}^{\infty} \sin \frac{n\pi x}{L} \left(b_n \cos \frac{cn\pi t}{L} + b_n^* \sin \frac{cn\pi t}{L} \right) \tag{4.5}$$

which satisfies the PDE and the boundary conditions. The coefficients of b_n and b_n^* are determined from the initial conditions $u(x,0) = f(x)$ and $u_t(x,0) = g(x)$. The question is how and why we can do it. In this chapter, we will present a systematical discussion about how to generate those normal modes and how to obtain b_n and b_n^* from initial conditions.

4.1 Orthogonal functions

Orthogonal functions are similar to orthogonal basis in the R^n space in linear algebra. Examples and applications include Fourier series, orthogonal polynomials, approximation theory and methods, and many more. One of notable applications is that we can expand functions in terms of orthogonal functions. Orthogonal functions are also intensively utilized in computational mathematics as approximation tools.

In the R^n space that is composed of all column vectors with n components, the simplest orthogonal basis are $\{\mathbf{e}_i\}_{i=1}^{n}$. For example, if $n = 3$. we have

$$\mathbf{e}_1 = \begin{pmatrix} 1 \\ 0 \\ 0 \end{pmatrix}, \quad \mathbf{e}_2 = \begin{pmatrix} 0 \\ 1 \\ 0 \end{pmatrix}, \quad \mathbf{e}_3 = \begin{pmatrix} 0 \\ 0 \\ 1 \end{pmatrix},$$

which satisfies

$$(\mathbf{e}_i, \mathbf{e}_j) = \mathbf{e}_i^T \mathbf{e}_j = \begin{cases} 1 & \text{if } i = j, \\ 0 & \text{if } i \neq j, \end{cases}$$

where $\mathbf{e}_i^T \mathbf{e}_j$ is the sum of the products of corresponding components of $\mathbf{e}_i$ and $\mathbf{e}_j$, which is the Euclidian inner product of the two vectors. For any vector $\mathbf{a} =$

$[a_1, a_2, a_3]^T$, we have $\mathbf{a} = a_1 \mathbf{e}_1 + a_2 \mathbf{e}_2 + a_3 \mathbf{e}_3$. If $\mathbf{b} = \{b_i\}$ is a vector, then the inner product of a and b is defined as $(a, b) = \sum_{i=1}^{3} a_i b_i$.

There are other orthogonal basis in R^3, for example,

$$\tilde{\mathbf{e}}_1 = \begin{pmatrix} 1 \\ 0 \\ 0 \end{pmatrix}, \quad \tilde{\mathbf{e}}_2 = \begin{pmatrix} 0 \\ -\frac{1}{\sqrt{2}} \\ \frac{1}{\sqrt{2}} \end{pmatrix}, \quad \tilde{\mathbf{e}}_3 = \begin{pmatrix} 0 \\ \frac{1}{\sqrt{2}} \\ \frac{1}{\sqrt{2}} \end{pmatrix},$$

also form a normalized orthogonal basis since $\tilde{\mathbf{e}}_i, i = 1, 2, 3$, have a unit length in the Euclidian norm. How do we express any vector in terms of $\{\tilde{\mathbf{e}}_i\}$? It is easy to check the following expressions,

$$\mathbf{a} = \alpha_1 \tilde{\mathbf{e}}_1 + \alpha_2 \tilde{\mathbf{e}}_2 + \alpha_3 \tilde{\mathbf{e}}_3, \qquad \alpha_i = \frac{(\mathbf{a}, \tilde{\mathbf{e}}_i)}{(\tilde{\mathbf{e}}_i, \tilde{\mathbf{e}}_i)} = (\mathbf{a}, \tilde{\mathbf{e}}_i).$$

Similar to the R^n space, we define a functional space which is a set of functions that has operations. All square integrable functions in an interval (a, b) form a linear space, called the $L^2(a, b)$ space,

$$L^2(a, b) = \left\{ f(x), \int_a^b |f(x)|^2 dx < \infty \right\}. \tag{4.6}$$

It is a linear space because if $f(x) \in L^2(a, b)$ and $g(x) \in L^2(a, b)$, then their linear combination $w(x) = \alpha f(x) + \beta g(x)$ is also in $L^2(a, b)$ for any constant α and β. In $L^2(a, b)$ we can define an **inner product** similar to that in R^n space as

$$(f, g) = \int_a^b f(x) \bar{g}(x) dx, \tag{4.7}$$

where $\bar{g}(x) = g(x)$ in the real number space and is the conjugate of $g(x)$ is the complex number space. For example, if $f(x) = e^x + i \sin x$, then $\bar{f}(x) = e^x - i \sin x$, where $i = \sqrt{-1}$ is the imaginary unit. We call $f(x)$ and $g(x)$ *orthogonal* (similar to perpendicular in Euclidean geometry) in $L^2(a, b)$ if $(f, g) = 0$.

Example 4.1. *Let $f(x) = 1$ and $g(x) = \sin x$. Are the two functions are orthogonal in $(0, 2\pi)$? We check the inner product*

$$(f, g) = \int_0^{2\pi} f(x) \bar{g}(x) dx = \int_0^{2\pi} \sin x \, dx = 0.$$

Thus, the two functions are orthogonal in the interval $(0, 2\pi)$ or any interval of length 2π, but it is not orthogonal in the interval $(0, \pi)$.

4.1. Orthogonal functions

The norm of a function $f(x)$ in $L^2(a,b)$ in terms of the $L^2(a,b)$ inner product is defined as

$$\|f\|_2 = \|f\|_{L^2} = \sqrt{(f,f)} = \sqrt{\int_a^b |f(x)|^2 dx}. \tag{4.8}$$

Often the subscript is omitted if there is no confusion occurs, that is, $\|f\| = \|f\|_2$.

Example 4.2. *Given an interval $(a,b) = (0, 2\pi)$, find the $L^2(a,b)$ norm of $f(x) = 1$ and $g(x) = \cos x$. According to the definition, we calculate,*

$$\|f\|_2 = \sqrt{\int_0^{2\pi} |f(x)|^2 dx} = \sqrt{2\pi};$$

$$\|g\|_2 = \sqrt{\int_0^{2\pi} |g(x)|^2 dx} = \sqrt{\int_0^{2\pi} \cos^2 x \, dx} = \sqrt{\pi}.$$

Example 4.3. *We redo the computation but with a different interval $(a,b) = (0, \pi)$.*

$$\|f\|_2 = \sqrt{\int_0^{\pi} |f(x)|^2 dx} = \sqrt{\pi};$$

$$\|g\|_2 = \sqrt{\int_0^{\pi} \cos^2 x \, dx} = \sqrt{\frac{\pi}{2}}.$$

Note that there are many different norms, for example,

$$\|f\|_1 = \int_a^b |f(x)| dx, \qquad \|f\|_\infty = \max_{0 \le x \le 2\pi} |f(x)|, \tag{4.9}$$

which are not associated with the $L^2(a,b)$ space. In fact, all these norms can be put in a uniform form

$$\|f\|_p = \left(\int_a^b |f(x)|^p dx \right)^{1/p} \tag{4.10}$$

for any $p > 0$. It can be shown that $\|f\|_\infty = \lim_{p \to \infty} \|f\|_p$. Note that only when $p = 2$, the norm is differentiable and it is why the L^2 norm has the most useful applications.

There are more than one ways to define an inner product, so is the related norm. An inner product is a special functional[2] of two arguments that satisfies

---
[2] A function whose arguments are functions.

- $(f, g) = \overline{(g, f)}$
- $(\alpha f + \beta g, h) = \alpha(f, g) + \beta(g, h)$ for any scalars α and β.

A norm is also a functional that should satisfy

- $\|f\| \geq 0$ and $\|f\| = 0$ if and only if $f(x) = 0$, or $\int_a^b f^2(x)dx = 0$;
- $\|\alpha f\| = |\alpha|\|f\|$ for any scalar α;
- $\|f + g\| \leq \|f\| + \|g\|$ which is called the triangle inequality.

All these statements are true in the R^n space. The famous Cauchy-Schwartz inequality is also true in $L^2(a,b)$ space, that is

$$|(f, g)| \leq \|f\|_2 \|g\|_2, \quad \text{or}$$

$$\left| \int_a^b f(x)g(x)dx \right| \leq \sqrt{\int_a^b f(x)^2 dx} \sqrt{\int_a^b g(x)^2 dx}. \quad (4.11)$$

Particularly, if we take $g(x) = 1$, we get

$$\left| \int_a^b f(x)dx \right|^2 \leq (b-a) \int_a^b f(x)^2 dx. \quad (4.12)$$

An example of different inner product is a weighted inner product. Let $w(x)$ be a non-negative function such that $w(x) \geq 0$ and $\int_a^b w(x)dx > 0$. The weighted inner product of $f(x)$ and $g(x)$ is defined as

$$(f, g)_w = \int_a^b f(x)\overline{g(x)}w(x)dx. \quad (4.13)$$

The function $f(x)$ and $g(x)$ are orthogonal with respect to $w(x)$ on (a,b) if

$$(f, g)_w = \int_a^b f(x)\overline{g(x)}w(x)dx = 0. \quad (4.14)$$

The corresponding norm is then

$$\|f\|_w = \sqrt{(f, f)_w} = \sqrt{\int_a^b w(x)|f(x)|^2 dx}. \quad (4.15)$$

We will see an applications of weighted inner products and norms for partial differential equations in polar and spherical coordinates for which $w(r) = r$.

Example 4.4. *Find the parameter a such that the two functions $f(x) = 1 + ax$ and $g(x) = \sin x$ are orthogonal with respect to the weight function $w(x) = x$ in $(0, \pi)$.*

Find the $L_w^2(0, \pi)$ norms of $f(x)$ and $g(x)$. Also find the two normalized orthogonal functions.

Solution: With some calculations or using the Maple, we get $\int_0^\pi (1+ax) x \sin x \, dx = \pi(1+a)$. To make the two functions orthogonal with respect to $w(x) = x$, we conclude that $a = -1$. Next, we compute

$$\|f\|_w = \sqrt{\int_0^\pi (1-x)^2 \, x \, dx} = \frac{\pi}{\sqrt{12}} \sqrt{3\pi^2 - 8\pi + 6},$$

$$\|g\|_w = \sqrt{\int_0^\pi \sin^2 x \cdot x \, dx} = \frac{\pi}{2}.$$

The two normalized orthogonal functions are

$$\bar{f}(x) = \frac{\sqrt{12}}{\pi\sqrt{3\pi^2 - 8\pi + 6}} (1-x), \quad \text{and} \quad \bar{g}(x) = \frac{2}{\pi} \sin x.$$

4.2 Function expansions in terms of orthogonal sets

We can represent or approximate a function $f(x)$ in terms of a set of orthogonal functions under some conditions.

Definition 4.1. *Let $f_1(x), f_2(x), \cdots, f_n(x), \cdots$ be a set of functions in $L^2(a,b)$, which can also be denoted as $\{f_n(x)\}_{n=1}^\infty$. It is called an orthogonal set if $(f_i, f_j) = 0$ as long as $i \neq j$ for all i and j's. The orthogonal set is called a **normalized** orthogonal set if $\|f_i\| = 1$ for all i's.*

Example 4.5. *The following functions,*

$$f_1(x) = \sin x, f_2(x) = \sin 2x, f_3(x) = \sin 3x, \cdots, f_n(x) = \sin nx, \cdots,$$

or $\{\sin nx\}_{n=1}^\infty$ is an orthogonal set in $L^2(-\pi, \pi)$.

Proof: If $m \neq n$, we can verify that

$$\int_{-\pi}^{\pi} \sin nx \sin mx \, dx = \int_{-\pi}^{\pi} -\frac{1}{2} \Big(\cos(m+n)x - \cos(m-n)x \Big) dx$$

$$= -\frac{1}{2} \left(\frac{\sin(m+n)x}{m+n} \bigg|_{-\pi}^{\pi} + \frac{\sin(m-n)x}{m-n} \bigg|_{-\pi}^{\pi} \right) = 0.$$

Note that if $m = n$, then we have

$$\int_{-\pi}^{\pi} \sin^2 nx\, dx = \int_{-\pi}^{\pi} \frac{1 - \cos 2nx}{2}\, dx = \pi. \tag{4.16}$$

Thus, we have $\|f_n\| = \sqrt{\pi}$. The new orthogonal set $\{\hat{f}_n(x)\} = \{f_n(x)/\sqrt{\pi}\}$ is a normalized orthogonal set.

Note also that the above discussions are true for any interval $(a, a + 2\pi)$ of length of 2π since $\sin nx$, $n \neq 0$, is a periodic function of period 2π.

4.2.1 Approximating functions using an orthogonal set

We can expand a function $f(x)$ using an orthogonal set of functions $\{f_n(x)\}$ that has a finite or infinite number of terms literally as

$$f(x) \sim \sum_{n=1}^{N} a_n f_n(x); \qquad f(x) \sim \sum_{n=1}^{\infty} a_n f_n(x). \tag{4.17}$$

While we can always do this, the left and right hand sides of above may not be the same, and that is why we use the '$\sim$' sign. To find out the coefficients $\{a_n\}_{n=1}^{\infty}$, we assume that the equal sign holds and apply the inner product of the above with a function $f_m(x)$ to get

$$\Big(f(x),\ f_m(x)\Big) = \Big(\sum_{n=1}^{\infty} a_n f_n(x),\ f_m(x)\Big) = \sum_{n=1}^{\infty} a_n \Big(f_n(x),\ f_m(x)\Big).$$

Since $\{f_n(x)\}_{n=1}$ is an orthogonal set, the right hand side terms are zeros except the m-th term, that is

$$\Big(f(x),\ f_m(x)\Big) = a_m \Big(f_m(x),\ f_m(x)\Big) \implies a_m = \frac{\big(f(x),\ f_m\big)}{\big(f_m(x),\ f_m(x)\big)}. \tag{4.18}$$

Expansion of $f(x)$ in terms of an orthogonal set $\{\phi_i(x)\}_{i=1}^{N}$ on an interval (a, b), where N can be ∞.

$$f(x) \sim \sum_{n=1}^{N} a_n \phi_n(x), \qquad a_n = \frac{\big(f(x),\ \phi_n(x)\big)}{\big(\phi_n(x),\ \phi_n(x)\big)} = \frac{\int_a^b f(x)\phi_n(x)\,dx}{\int_a^b \phi_n^2(x)\,dx}.$$

Example 4.6. *Expand $f(x) = x$ in terms of $\{\sin nx\}$ on $(-\pi, \pi)$.*

4.2. Function expansions in terms of orthogonal sets

We know that $\{\sin nx\}$ is an orthogonal set on $(-\pi, \pi)$. The coefficient a_n is

$$a_n = \frac{\int_{-\pi}^{\pi} x \sin nx \, dx}{\int_{-\pi}^{\pi} \sin^2 nx \, dx} = \frac{1}{\pi}\left(-\frac{x\cos nx}{n}\bigg|_{-\pi}^{\pi} + \int_{-\pi}^{\pi} \frac{\cos nx}{n} dx\right)$$

$$= -\frac{2}{\pi}\frac{\cos n\pi}{n}.$$

The expansion then is

$$x \sim 2\sin x - \sin 2x + \frac{2}{3}\sin 3x - \cdots = \sum_{n=1}^{\infty} \frac{2(-1)^{n-1}}{n\pi}\sin nx.$$

From the Fourier series theory, we know that the equality sign holds for this case at any x in $(-\pi, \pi)$ but not at two end points $-\pi$ and π.

Example 4.7. *Expand $f(x) = x^2$ in terms of $\{\sin nx\}$ on $(-\pi, \pi)$.*

It is easy to check that $a_n = 0$ for all n's. This is because we have

$$a_n = \frac{\int_{-\pi}^{\pi} x^2 \sin nx \, dx}{\int_{-\pi}^{\pi} \sin^2 nx \, dx} = 0.$$

The integrand is an odd function whose integral in a symmetric interval is always 0. Such an expansion is meaningless. This is because the function $f(x) = x^2$ does not share much properties of the orthogonal set $\{\sin nx\}$ on $(-\pi, \pi)$.

We call the orthogonal set $\{\sin nx\}$ on $(-\pi, \pi)$ is incomplete or a subset in the space $L^2(\pi, \pi)$. In Figure 4.1, we show a diagram among function sets in $L^2(a, b)$. The sets $\{\sin nx\}$ and $\{\cos nx\}$ are subsets of $L^2(-\pi, \pi)$. While $\{\sin nx\}$ or $\{\cos nx\}$ is not complete in $L^2(-\pi, \pi)$ meaning that not all the functions in the space can have meaningful expansions by the orthogonal sets, they are complete in some smaller spaces if additional conditions are imposed such as some kind of boundary conditions, even or odd functions *etc*.

It is easy to check that the set $\{\cos nx\}_{n=0}^{\infty}$ is also an orthogonal on $(-\pi, \pi)$. Note that this set includes $f(x) = 1$ when $n = 0$. We can expand $f(x) = x^2$ in terms of $\{\cos nx\}_{n=0}^{\infty}$. But it is meaningless to expand $f(x) = x$ in terms of $\{\cos nx\}_{n=0}^{\infty}$. However, if we put the two orthogonal sets together to form a new set $\{1, \cos nx, \sin nx\}_{n=1}^{\infty}$, then we can show that the new set is another orthogonal set since $\int_{-\pi}^{\pi} \sin mx \cos nx = 0$ for any m and n. Any function $f(x)$ in $L^2(-\pi, \pi)$ can be expanded by the orthogonal set,

$$f(x) \sim \frac{a_0}{2} + \sum_{n=1}^{\infty}(a_n \cos nx + b_n \sin nx), \qquad (4.19)$$

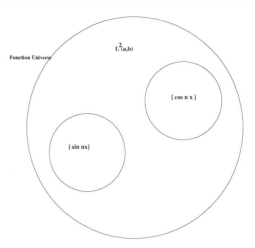

Figure 4.1. *A diagram of orthogonal function spaces in $L^2(a,b)$. If we regard all functions as a universe because no one can count them, then $L^2(a,b)$ is a complete subset, called a Hilbert space since an inner product is defined. $L^2(a,b)$ is complete meaning that any Cauchy sequence will converge to a function in $L^2(a,b)$. The orthogonal set $\{\sin nx\}$ and $\{\cos nx\}$ are subsets of $L^2(0,\pi)$.*

where

$$
\begin{aligned}
a_n &= \frac{1}{\pi}\int_{-\pi}^{\pi} f(x)\cos nx\, dx, & n &= 0,1,\cdots; \\
b_n &= \frac{1}{\pi}\int_{-\pi}^{\pi} f(x)\sin nx\, dx, & n &= 1,2,\cdots
\end{aligned}
\quad (4.20)
$$

This is called a Fourier series of $f(x)$ on $(-\pi,\pi)$. The reason to have a factor $\frac{1}{2}$ for a_0 is that we can have a uniform formula for $\{a_n\}_{n=0}^{\infty}$.

4.3 Sturm-Liouville eigenvalue problems

Sturm-Liouville (S-L) eigenvalue problems provide a way of generating orthogonal functions that have some special properties. One example is the S-L eigenvalue problem obtained from the method of separation of variables for one-dimensional wave equations $u_{tt} = c^2 u_{xx}$ in the domain $(0,L)$ with homogeneous boundary conditions $u(0,t) = u(L,t) = 0$. The Sturm-Liouville eigenvalue problem would lead to a set of orthogonal functions $\left\{\sin\frac{n\pi x}{L}\right\}$. For any function $f(x) \in L^2(0,L)$ with $f(0) = 0$ and $f(L) = 0$, we can have a meaningful expansion of $f(x)$ in terms of the orthogonal functions $\left\{\sin\frac{n\pi x}{L}\right\}$.

4.3. Sturm-Liouville eigenvalue problems

Here we discuss Sturm-Liouville problems that have the following form

$$(p(x)y'(x))' + q(x)y(x) = f(x), \quad a < x < b \tag{4.21}$$

with boundary conditions (BC) at $x = a$ and $x = b$. Take $x = a$ for example, three types of linear boundary conditions are often used.

(a) The solution is prescribed, that is, $y(a) = \alpha$ is known. It is called a Dirichlet BC.

(b) The derivative of the solution is prescribed, that is, $y'(a) = \beta$ is known. It is called a Neumann BC. Note that the solution is unknown at $x = a$.

(c) The BC is prescribed as $\alpha y(a) + \beta y'(a) = \gamma$ with $\beta \neq 0$. It is called a Robin or mixed BC.

We can write down a uniform form of the three boundary conditions at the two ends as

- $c_1 y(a) + c_2 y'(a) = b_1, \quad c_1^2 + c_2^2 \neq 0;$
- $d_1 y(b) + d_2 y'(b) = b_1, \quad d_1^2 + d_2^2 \neq 0.$

The notation $c_1^2 + c_2^2 \neq 0$ means that c_1 and c_2 cannot be both zero simultaneously. The ordinary differential equation (ODE) is called a self-adjoint ODE. Note that $p(x)y''(x) + w(x)y'(x) + q(x)y(x) = f(x)$ is not a self-adjoint ODE unless it can be transformed to the standard form $(\bar{p}(x)y'(x))' + \bar{q}(x)y(x) = f(x)$.

A Sturm-Liouville problem will have a unique solution if both $p(x)$, $q(x)$, and $f(x)$ are continuous,[3] and $p(x) \geq p_0 > 0$ and $q(x) \leq 0$ with suitable boundary conditions, for example, a Dirichlet BC at one of two ends. However, here we are more interested in problems below that have multiple solutions

$$\Big(p(x)y'(x)\Big)' + \Big(q(x) + \lambda r(x)\Big) y(x) = 0, \quad a < x < b,$$
$$c_1 y(a) + c_2 y'(a) = 0, \quad c_1^2 + c_2^2 \neq 0, \tag{4.22}$$
$$d_1 y(b) + d_2 y'(b) = 0, \quad d_1^2 + d_2^2 \neq 0,$$

with both $y(x)$ and λ being unknowns. Such problems are called Sturm-Liouville eigenvalue problems. Note that the ODE and the boundary conditions are all homogeneous and $r(x)$ is a weight function.

---

[3] These conditions can be lessened in high level mathematics.

Apparently $y(x) = 0$ is a solution, called a *trivial solution*. We can find some λ such that the problem has non-trivial solutions. In a Sturm-Liouville eigenvalue problem, we want to find both an eigenvalue λ, and a corresponding eigenfunction $y_\lambda(x) \neq 0$ that satisfies both of the ODE and the boundary conditions. We call such $((\lambda, y_\lambda(x))$ an eigenpair of the S-L eigenvalue problem.

Example 4.8. *Solve the eigenvalue problem*

$$y'' + \lambda y = 0, \quad 0 < x < \pi,$$

$$y(0) = y(\pi) = 0.$$

Solution: In this example $p(x) = 1$, $q(x) = 0$, and $r(x) = 1$. The roots of the characteristic polynomial of the ODE are $\pm\sqrt{-\lambda}$. If $\lambda < 0$, then the solution is

$$y(x) = C_1 e^{\sqrt{-\lambda}x} + C_2 e^{-\sqrt{-\lambda}x}.$$

Plugging the boundary conditions $y(0) = y(\pi) = 0$ into the above, we get

$$C_1 + C_2 = 0, \qquad C_1 e^{\sqrt{-\lambda}\pi} + C_2 e^{-\sqrt{-\lambda}\pi} = 0.$$

The only solution is $C_1 = 0$ and $C_2 = 0$, which leads to a trivial solution $y(x) = 0$. Similarly if $\lambda = 0$, then $y(x) = C_1 + C_2 x$, which again leads to $y(x) = 0$ using the boundary conditions.

However, if $\lambda > 0$, then the general solution is

$$y(x) = C_1 \cos\sqrt{\lambda}x + C_2 \sin\sqrt{\lambda}x.$$

The boundary condition $y(0) = 0$ leads to $C_1 = 0$. Thus $y(x) = C_2 \sin\sqrt{\lambda}x$. The second boundary condition $y(\pi) = 0$ leads to $C_2 \sin\sqrt{\lambda}\pi = 0$. When does $\sin(x) = 0$? We know that $x = 0$, or π, or 2π, or $\cdots$, and so on, which leads to $x = n\pi$. Thus, we get

$$\sqrt{\lambda}\pi = n\pi \longrightarrow \lambda = n^2, \quad n = 1, 2, \cdots.$$

Note that $n = 0$ leads to a trivial solution and should be discarded. The solutions to the eigenvalue problem are

$$\lambda_n = n^2, \qquad y_n(x) = \sin nx, \qquad n = 1, 2, \cdots.$$

Usually, we do not include constant C_2 term since eigenfunctions can differ by constants. Note also that the eigenfunctions $\{\sin nx\}$ is an orthogonal set in $(0, \pi)$.

4.3. Sturm-Liouville eigenvalue problems

Class practice

Solve the eigenvalue problem

$$y'' + \lambda y = 0, \quad 0 < x < 1,$$
$$y(0) = y(1) = 0.$$

The solution is $\lambda_n = (n\pi)^2$ and $y_n(x) = \sin n\pi x$, for $n = 1, 2, \cdots$.

4.3.1 Regular and singular Sturm-Liouville eigenvalue problems

Consider again a Sturm-Liouville eigenvalue problem,

$$\begin{aligned}
(p(x)y'(x))' + (q(x) + \lambda r(x))y(x) &= 0, \quad a < x < b, \\
c_1 y(a) + c_2 y'(a) &= 0, \quad (c_1)^2 + (c_2)^2 \neq 0, \\
d_1 y(b) + d_2 y'(b) &= 0, \quad (d_1)^2 + (d_2)^2 \neq 0.
\end{aligned} \quad (4.23)$$

Mathematically we require

1. $p(x)$, $q(x)$ and $r(x)$ are all continuous, or $p, q, r \in C[a, b]$ for short.

2. $p(x) \geq p_0 > 0$ and $r(x) \geq 0$ for $a \leq x \leq b$,

where p_0 is a positive constant. Such a problem is called a *regular Sturm-Liouville problem*. For a regular Sturm-Liouville eigenvalue problem, the eigenfunctions are all continuous and bounded in (a, b). From advanced differential equations theories, we require $p(x)$ is continuous and non-zero in (a, b) so that the ordinary differential equation remains to be a second order ODE. If the conditions, especially, the condition on $p(x)$ is violated, we called the Sturm-Liouville eigenvalue problem, a *singular* problem. Below are some examples:

$$y'' + \lambda y = 0, \quad -1 < x < 1, \quad \text{regular};$$
$$(xy')' + \lambda y = 0, \quad -1 < x < 1, \quad \text{singular at } x = 0;$$
$$((1 - x^2)y')' + \lambda y = 0, \quad -1 < x < 1, \quad \text{singular at } x = \pm 1.$$

Sometime, we need some effort to re-write a problem to have a standard Sturm-Liouville eigenvalue form to judge whether the problem is regular or singular.

Example 4.9. $x^2 y'' + 2xy' + \lambda y = 0$ can be written as $(x^2 y')' + 2xy' + \lambda y - 2xy' = 0$, which is $(x^2 y')' + \lambda y = 0$. The eigenvalue problem is a regular in an interval (a, b)

that does not contain the original, and that the zero is not one of two end points. Otherwise it would be singular.

Example 4.10. *We can divide by x^2 for $xy'' - y' + \lambda xy = 0$ to get $\frac{1}{x}y'' - \frac{1}{x^2}y' + \lambda y = 0$ which is $\left(\frac{1}{x}y'\right)' + \lambda y = 0$, which is a standard S-L eigenvalue problem. The discussion in the previous example about whether the problem is regular or singular also applies to this S-L eigenvalue problem.*

Below we present an example for a different boundary condition, a Neumann boundary condition at $x = b$.

Example 4.11. *Solve the eigenvalue problem*

$$y'' + \lambda y = 0, \quad 0 < x < \pi,$$

$$y(0) = 0, \quad y'(\pi) = 0.$$

Solution: From previous examples, we know that the solution should be $y(x) = C_2 \sin \sqrt{\lambda} x$. Thus the derivative is $y'(x) = C_2 \sqrt{\lambda} \cos \sqrt{\lambda} x$. From $y'(\pi) = 0$ we get $y'(\pi) = \cos \sqrt{\lambda} \pi = 0$. Thus, the eigenvalues are

$$\sqrt{\lambda} \pi = \left(\frac{1}{2} + n\right) \pi, \quad n = 0, 1, 2, \cdots, \quad \Longrightarrow \quad \lambda_n = \left(\frac{1}{2} + n\right)^2,$$

$$y_n(x) = \sin\left(\frac{1}{2} + n\right) x.$$

Question: Can we take $n = -1, n = -2, \cdots$?

The eigenfunctions $\left\{\sin(\frac{1}{2} + n)x\right\}_{n=0}^{\infty}$ form an orthogonal set that can be used to solve the wave equations $u_{tt} = c^2 u_{xx}$ with the boundary condition $u(0,t) = 0$ and $\frac{\partial u}{\partial x}(\pi, t) = 0$ on the interval $(0, \pi)$.

Example 4.12. *Solve the eigenvalue problem with a mixed boundary condition (also called a Robin BC)*

$$y'' + \lambda y = 0, \quad 0 < x < 1,$$

$$y'(0) = 0, \quad y(1) + y'(1) = 0.$$

Solution: From previous discussions, we know that the solution should have

4.3. Sturm-Liouville eigenvalue problems

the form

$$y(x) = y(x) = C_1 \cos(\sqrt{\lambda}x) + C_2 \sin(\sqrt{\lambda}x),$$
$$y'(x) = -\sqrt{\lambda}\, C_1 \sin(\sqrt{\lambda}x) + \sqrt{\lambda}\, C_2 \cos(\sqrt{\lambda}x).$$

From $y'(0) = 0$, we conclude that $C_2 = 0$ since $\sqrt{\lambda} = 0$ implies a trivial solution $y = 0$. From the mixed boundary condition we have

$$C_1 \left(\cos \sqrt{\lambda} - \sqrt{\lambda} \sin \sqrt{\lambda} \right) = 0, \quad \text{or} \quad \cos \sqrt{\lambda} - \sqrt{\lambda} \sin \sqrt{\lambda} = 0, \quad \Longrightarrow \quad \cot \sqrt{\lambda} = \sqrt{\lambda}.$$

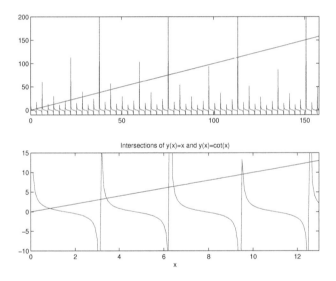

Figure 4.2. *Plots of $y = x$ and $y = \cot x$. The intersections are the squares of eigenvalues of λ_n. Note that round-off errors from the computer and the effect of the singularities of $\cot x$ at $k\pi$, $k = 1, 2, \cdots$, are visible.*

There is no closed form for the eigenvalues, which are zeros of a non-linear equation. But we do know that the squares of the eigenvalues are the intersections of the graphs of $y = x$ and $y = \cot x$ in the half plane $x > 0$. There are infinite number of intersections in the first quadrant at: $\alpha_1 = 0.86 \cdots$, $\alpha_2 = 3.43 \cdots$, $\alpha_3 = 6.44 \cdots$. The eigenvalues are $\lambda_n = \alpha_n^2$, and the eigenfunctions are $y_n(x) = \cos \lambda_n x$. In Figure 4.2, we show two plots of $y = x$ and $y = \cot x$. The intersections are $\alpha_n = \sqrt{\lambda_n}$.

4.4 Theory and applications of Sturm-Liouville eigenvalue problems

For a regular Sturm-Liouville eigenvalue problem,

$$\Big(p(x)y'(x)\Big)' + \Big(q(x) + \lambda r(x)\Big)y(x) = 0, \quad a < x < b,$$
$$c_1 y(a) + c_2 y'(a) = 0, \quad c_1^2 + c_2^2 \neq 0, \quad (4.24)$$
$$d_1 y(b) + d_2 y'(b) = 0, \quad d_1^2 + d_2^2 \neq 0.$$

Assume that all $p(x), q(x), r(x) \in C(a,b)$ are real functions, $p(x) \geq p_0 > 0$, $r(x) \geq 0$, $\int_a^b r(x) > 0$. Then we have the following theorem.

Theorem 4.2.

1. *There are infinite number of eigenvalues, which are all real numbers. We can arrange the eigenvalues as*

$$\lambda_1 < \lambda_2 < \cdots < \lambda_n < \cdots, \quad \lim_{n \to \infty} \lambda_n = \infty. \quad (4.25)$$

   Furthermore, If $q(x) \leq 0$, then all the eigenvalues are positive $\lambda_n > 0$ for $n = 1, 2, \cdots$.

2. *The eigenfunctions $y_1(x), y_2(x), \cdots, y_n(x), \cdots$, form an orthogonal set with respect to the weight function $r(x)$ on (a,b), that is*

$$\int_a^b y_m(x) y_n(x) r(x) dx = 0, \quad \text{if} \quad m \neq n. \quad (4.26)$$

3. *For any function $u(x) \in L_r^2(a,b)$ that satisfies the same boundary condition, $u(x)$ can be expanded in terms of the orthogonal set $\{y_n(x)\}_{n=1}^\infty$, that is,*

$$u(x) = \sum_{n=1}^\infty A_n y_n(x) \quad \text{with} \quad A_n = \frac{\int_a^b u(x) y_n(x) r(x) dx}{\int_a^b y_n^2(x) r(x) dx}. \quad (4.27)$$

Sketch of the proof of the orthogonality: Let $y_k(x)$ and $y_j(x)$ be two distinct eigenfunctions corresponding to the eigenvalues of λ_k and λ_j, respectively,

4.4. Theory and applications of Sturm-Liouville eigenvalue problems

that is,

$$\left(py_j'\right)' + \left(q + \lambda_j r\right) y_j = 0, \tag{4.28}$$

$$\left(py_k'\right)' + \left(q + \lambda_k r\right) y_k = 0. \tag{4.29}$$

We multiply (4.29) by $y_j(x)$ and multiply (4.28) by $y_k(x)$; and then subtract the two to get

$$y_k \left(py_j'\right)' - y_j \left(py_k'\right)' + (\lambda_j - \lambda_k) r y_j y_k = 0. \tag{4.30}$$

Integrating above from a to b leads to

$$(\lambda_j - \lambda_k) \int_a^b r y_j y_k dx = \int_a^b y_k \left((py_j')' - y_j(py_k')'\right) dx.$$

Applying integration by parts to the right hand side and carrying out some manipulations, we get

$$(\lambda_j - \lambda_k) \int_a^b r y_j y_k dx = p(b) y_j'(b) y_k(b) - p(a) y_j(b) y_k'(b)$$
$$- p(a) y_j'(a) y_k(a) + p(a) y_j(a) y_k'(a).$$

From the boundary condition at $x = a$ we have

$$\begin{bmatrix} y_j(a) & y_j'(a) \\ y_k(a) & y_k'(a) \end{bmatrix} \begin{bmatrix} c_1 \\ c_2 \end{bmatrix} = \begin{bmatrix} 0 \\ 0 \end{bmatrix}.$$

Since $c_1^2 + c_2^2 \neq 0$, we must have that the determinant of the 2 x 2 coefficient matrix must be zero, that is, $y_j'(a) y_k(a) - y_j(a) y_k'(a) = 0$. Since $p(a) \neq 0$, we conclude that

$$p(a) y_j'(a) y_k(a) - p(a) y_j(a) y_k'(a) = 0.$$

By the same derivation at $x = b$, we also have

$$p(b) y_j'(b) y_k(b) - p(b) y_j(b) y_k'(b) = 0.$$

Thus, we have $(\lambda_j - \lambda_k) \int_a^b r y_j y_k dx = 0$ and since $\lambda_j \neq \lambda_k$, we conclude that $\int_a^b r y_j y_k dx = 0$. This completes the proof.

4.5 Application of the S-L eigenvalue theory and orthogonal expansions

Let us revisit the initial and boundary value problem of one-dimensional wave equations,

$$\frac{\partial^2 u}{\partial t^2} = c^2 \frac{\partial^2 u}{\partial x^2}, \quad 0 < x < L,$$

$$u(0,t) = 0, \quad u(L,t) = 0,$$

$$u(0,t) = f(x), \quad \frac{\partial u}{\partial t}(x,0) = g(x), \quad 0 < x < L,$$

for general $f(x)$ and $g(x)$. As discussed at the beginning of the chapter, the solution can be expressed as a superposition of normal mode solution,

$$u(x,t) = \sum_{n=0}^{\infty} \sin \frac{n\pi x}{L} \left(b_n \cos \frac{cn\pi t}{L} + b_n^* \sin \frac{cn\pi t}{L} \right) \tag{4.31}$$

that satisfies the PDE and the boundary conditions. The coefficients of b_n and b_n^* are determined from the initial conditions $u(x,0) = f(x)$ and $u_t(x,0) = g(x)$. Using the orthogonal expansion process, we have

$$u(x,0) = \sum_{n=1}^{\infty} b_n \sin \frac{n\pi x}{L}, \quad \Longrightarrow \quad b_n = \frac{2}{L} \int_0^L f(x) \sin \frac{n\pi x}{L} dx,$$

$$\frac{\partial u}{\partial t}(x,0) = \sum_{n=0}^{\infty} \sin \frac{n\pi x}{L} \left(-b_n \frac{cn\pi}{L} \sin \frac{cn\pi t}{L} + b_n^* \frac{cn\pi}{L} \cos \frac{cn\pi t}{L} \right),$$

$$\frac{\partial u}{\partial t}(x,0) = \sum_{n=1}^{\infty} \sin \frac{cn\pi t}{L} b_n^* \frac{cn\pi}{L} \quad \Longrightarrow \quad b_n^* = \frac{2}{cn\pi} \int_0^L g(x) \sin \frac{n\pi x}{L} dx.$$

Thus, the coefficients b_n's are the coefficients of the orthogonal expansions of $u(x,0) = f(x)$ in terms of the eigenfunctions; while the coefficients b_n^* are the coefficients of the orthogonal expansions of $u_t(x,0) = g(x)$ in terms of the eigenfunctions differed by some constants.

Solution to the 1D wave equation with homogeneous BCs

$$u(x,t) = \sum_{n=1}^{\infty} \sin \frac{n\pi x}{L} \left(b_n \cos \frac{cn\pi t}{L} + b_n^* \sin \frac{cn\pi t}{L} \right)$$

$$b_n = \frac{2}{L} \int_0^L f(x) \sin \frac{n\pi x}{L} dx, \quad b_n^* = \frac{2}{cn\pi} \int_0^L g(x) \sin \frac{n\pi x}{L} dx.$$

4.5. Application of the S-L eigenvalue theory and orthogonal expansions

Example 4.13. *Solve the wave equation,*

$$\frac{\partial^2 u}{\partial t^2} = \frac{\partial^2 u}{\partial x^2}, \quad 0 < x < 1,$$

$$u(0,t) = 0, \quad u(1,t) = 0,$$

$$u(x,0) = \begin{cases} 1 & \text{if } 0 \le x < \frac{1}{2}, \\ 0 & \text{if } \frac{1}{2} \le x \le 1, \end{cases} \quad \frac{\partial u}{\partial t}(x,0) = 0.$$

Solution: In this example, $c = 1$, $L = 1$, and $g(x) = 0$, we have $b_n^* = 0$ and

$$b_n = 2\int_0^{\frac{1}{2}} f(x)\sin n\pi x\, dx = 2\int_0^{\frac{1}{2}} \sin n\pi x\, dx = -\frac{2}{n\pi}\cos n\pi x \Big|_0^{\frac{1}{2}}$$

$$= -\frac{2}{n\pi}\left(\cos\frac{n\pi}{2} - 1\right) = \frac{2}{n\pi}\left(1 - \cos\frac{n\pi}{2}\right).$$

The solution to the wave equation is

$$u(x,t) = \sum_{n=1}^{\infty} \frac{2}{n\pi}\left(1 - \cos\frac{n\pi}{2}\right)\sin n\pi x \cos n\pi t.$$

We know that the series is convergent in the interval $(0, 1)$. In Figure 4.3, we show several plots of the partial sums defined as

$$S_n(x) = \sum_{n=1}^{N} \frac{2}{n\pi}\left(1 - \cos\frac{n\pi}{2}\right)\sin n\pi x \cos n\pi t. \tag{4.32}$$

with $N = 1$, $N = 5$, and $N = 75$ at $t = 0$. The series approximates the function $u(x, 0)$ well in the interior of continuous regions when N is large enough but oscillates at $x = 0$ as well as at the discontinuity $x = 1/2$, which is called the Gibb's phenomena. In the Maple file, one can use the animation feature to see the evolution of the solution with time t. It is interesting to observe how the discontinuity moves.

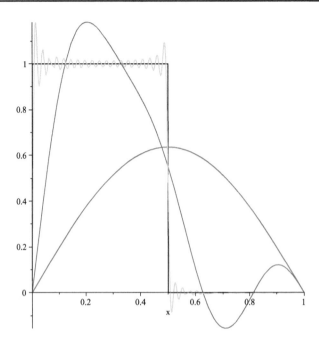

Figure 4.3. *Plots of the series approximations (partial sums) of the initial condition to the wave equation.*

Example 4.14. *Solve the 1D wave equation*

$$\frac{\partial^2 u}{\partial t^2} = c^2 \frac{\partial^2 u}{\partial x^2}, \quad 0 < x < 1,$$

$$u(0,t) = 0, \quad u(1,t) = 0,$$

$$u(x,0) = \sin \pi x - \frac{1}{2}\sin 2\pi x + \frac{1}{3}\sin 3\pi x, \quad \frac{\partial u}{\partial t}(x,0) = 0.$$

Solution: For this example, the initial conditions $u(x,0) = f(x)$ and $u_t(x,0) = g(x)$ are some normal modes and in the expansion forms already. Thus, the solution is a combination of normal mode solutions,

$$u(x,t) = \sin \pi x \cos \pi ct - \frac{1}{2}\sin 2\pi x \cos 2\pi ct + \frac{1}{3}\sin 3\pi x \cos 3\pi ct.$$

4.6 Series solution of 1D heat equations of initial and boundary value problems

We use the method of separation of variables to solve,

$$\frac{\partial u}{\partial t} = c^2 \frac{\partial^2 u}{\partial x^2}, \quad 0 < x < L,$$
$$u(0,t) = 0, \quad u(L,t) = 0,$$
$$u(x,0) = f(x), \quad 0 < x < L,$$

for a general $f(x) \in L^2(0,L)$. The PDE is called a one-dimensional heat equation. Note that there is only one initial condition. From the initial and boundary conditions, we should have $f(0) = u(0,0) = 0$ and $f(L) = u(L,0) = 0$. If these two conditions are satisfied, we call the initial and boundary conditions are consistent, which is not always true in some applications. The method of separation of variables includes the following steps.

Step 1: Let $u(x,t) = T(t)X(x)$ and we plug its partial derivatives into the original PDE so that we can separate variables. The homogeneous boundary conditions require $X(0) = X(L) = 0$. Differentiating with $u(x,t) = T(t)X(x)$ with t and x respectively, we get

$$\frac{\partial u}{\partial t} = T'(t)X(x); \quad \frac{\partial u}{\partial x} = T(t)X'(x), \quad \frac{\partial^2 u}{\partial x^2} = T(t)X''(x).$$

The 1D heat equation can be re-written as

$$T'(t)X(x) = c^2 T(t) X''(x) \implies \frac{T'(t)}{c^2 T(t)} = \frac{X''(x)}{X(x)} = -\lambda, \quad (4.33)$$

where λ is a constant for given x and t. This is because in the last equality, the left hand side is a function of t while the right hand side is a function of x, which is possible only both of them are a constant independent of t and x. We need to decide which is an eigenvalue problem that we can solve. Since we know the boundary condition for $X(x)$, naturally we should solve

$$\frac{X''(x)}{X(x)} = -\lambda \quad \text{or} \quad X''(x) + \lambda X(x) = 0, \quad X(0) = X(L) = 0 \quad (4.34)$$

first.

Step 2: Solve the eigenvalue problem. From the Sturm-Liouville eigenvalue theory, we know that $\lambda > 0$. Thus, the solution is

$$X''(x) = C_1 \cos \sqrt{\lambda} x + C_2 \sin \sqrt{\lambda} x.$$

From the boundary condition $X(0) = 0$, we get $C_1 = 0$. From the boundary condition $X(L) = 0$, we have

$$C_2 \sin \sqrt{\lambda} L = 0, \implies \sqrt{\lambda} L = n\pi, \quad n = 1, 2, \cdots,$$

since $C_2 \neq 0$ for non-trivial solutions. The eigenvalues and their corresponding eigenfunctions are

$$\lambda_n = \left(\frac{n\pi}{L}\right)^2, \quad X_n(x) = \sin \frac{n\pi x}{L}, \quad n = 1, 2, \cdots.$$

Next, we solve for $T(t)$ using

$$T'(t) + c^2 \lambda_n T(t) = 0, \tag{4.35}$$

with known $\lambda_n = \left(\frac{n\pi}{L}\right)^2$. The solution is (not an eigenvalue problem anymore since we have already known λ_n)

$$T_n(t) = b_n e^{-c^2 \lambda_n t} = b_n e^{-c^2 \left(\frac{n\pi}{L}\right)^2 t}.$$

Put $X_n(x)$ and $T_n(t)$ together, we get a normal mode solution

$$u_n(x, t) = b_n e^{-c^2 \left(\frac{n\pi}{L}\right)^2 t} \sin \frac{n\pi x}{L}, \tag{4.36}$$

which satisfy the PDE, the boundary conditions, but not the initial condition.

Step 3: Put all the normal mode solutions together to get the series solution. The coefficients are obtained from the orthogonal expansion of the initial condition.

The solution to the 1D heat equation can be written as

$$u(x, t) = \sum_{n=1}^{\infty} b_n \sin \frac{n\pi x}{L} e^{-c^2 \left(\frac{n\pi}{L}\right)^2 t} \tag{4.37}$$

which satisfies the PDE and the boundary conditions. The coefficients of b_n are determined from the initial conditions $u(x, 0)$,

$$u(x, 0) = \sum_{n=0}^{\infty} b_n \sin \frac{n\pi x}{L}, \implies b_n = \frac{2}{L} \int_0^L f(x) \sin \frac{n\pi x}{L} dx.$$

Series solution to the 1D heat equation is

$$u(x, t) = \sum_{n=1}^{\infty} b_n \sin \frac{n\pi x}{L} e^{-c^2 \left(\frac{n\pi}{L}\right)^2 t}, \quad b_n = \frac{2}{L} \int_0^L f(x) \sin \frac{n\pi x}{L} dx.$$

4.6. Series solution of 1D heat equations of initial and boundary value problems

Example 4.15. *Solve the 1D heat equation with homogeneous boundary conditions in the interval $(0, \pi)$ and the initial condition $u(x, 0) = 100$. What is the limit of $\lim_{t \to \infty} u(x, t)$?*

Solution: We use the formula above to find the coefficients of the series
$$b_n = \frac{2}{\pi} \int_0^\pi 100 \sin(nx) dx = \frac{200}{n\pi} \cos(nx) \Big|_0^\pi = \frac{200(1 - \cos(n\pi))}{n\pi}.$$
Thus, the solution is
$$u(x, t) = \sum_{n=1}^\infty \frac{200(1 - \cos(n\pi))}{n\pi} \sin(nx) e^{-n^2 t}.$$
Furthermore, we can easily show that $\lim_{t \to \infty} u(x, t) = 0$.

In Figure 4.4, we show plots of several partial sums of the initial condition with $N = 1$, $N = 5$, and $N = 175$. In the middle, the series approximate the function very well when N is large enough but oscillates at two end points, which is called the Gibb's phenomena. However, for heat equations, the oscillations will soon be dampened and the solution becomes smooth with the time. In the Maple file, one can use the animation feature to see the evolution of the solution.

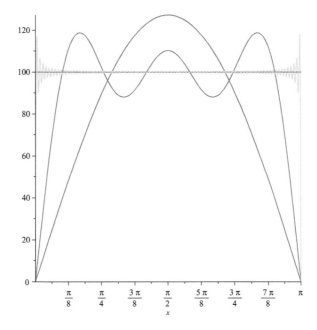

Figure 4.4. *Plots of the series approximations to the initial condition $u_0(x, 0) = 100$.*

4.7 Exercises

E4.1 Find all values of b such that the following two functions are orthogonal functions with respect to the weight function $w(x) = e^{-x}$ on the interval $0 < x < \pi$,
$$f_1(x) = \cos(bx), \quad f_2(x) = e^x.$$

E4.2 Given $\left\{\cos\dfrac{n\pi x}{p}\right\}_{n=0}^{N} = \left\{1, \cos\dfrac{\pi x}{p}, \cos\dfrac{2\pi x}{p}, \cdots, \cos\dfrac{i\pi x}{p}, \cdots, \cos\dfrac{N\pi x}{p}\right\}$.

(a) Show that the set forms an orthogonal set in $(-p, p)$. **Hint:** $\cos\alpha\cos\beta = \dfrac{1}{2}(\cos(\alpha+\beta) + \cos(\alpha-\beta))$.

(b) Find the L^2 norm $\|f\|_{L^2} = \sqrt{\int_{-p}^{p} f^2(x)dx}$ of $f(x) = 1$ ($n = 0$) and $f(x) = \cos\dfrac{n\pi x}{p}$ (other n).

(c) Find the orthogonal expansion of $f(x) = |x|$ in terms of the orthogonal set in $(-p, p)$.

Hint: $\displaystyle\int x\cos ax\, dx = \dfrac{x\sin ax}{a} + \dfrac{\cos ax}{a^2}.$

E4.3 Determine the constants a and b so that the functions 1, x, and $a + bx + x^2$ are orthogonal on $(-1, 1)$. Find the orthogonal function expansion of $\sin x$ or $\cos x$ (**choose one**), for example,
$$\sin x \sim \alpha_1 + \alpha_2 x + \alpha_3(a + bx + x^2)$$
after you have found a and b. Plot the function $\sin x$ and the approximation together.

E4.4 Let $\{\phi_n(x)\}_{n=1}^{\infty}$ be a set of orthogonal functions with respect to a weight function $w(x)$ in an interval (a, b).

(a) What does this mean?

(b) If $f(x) = \displaystyle\sum_{n=1}^{\infty} a_n\phi_n(x)$, then find the integral formula for a_n in terms of the appropriate functions.

E4.5 Find all the eigenvalues and eigenfunctions of the S-L. problem. Show all the cases and process.
$$y'' + \lambda y = 0, \quad 0 < x < \pi/2;$$
$$y(0) = 0, \quad y'(\pi/2) = 0$$

Also answer the following questions:

(a) Can the eigenvalues of regular S-L be complex numbers?

(b) Are there finite or infinite number of distinct eigenvalues?

(c) If $y_m(x)$ and $y_n(x)$ are two eigenfunctions corresponding to two different eigenvalues λ_m and λ_n, what is the result of $\int_0^{\pi/2} y_m(x)y_n(x)dx$?

E4.6 Expand $f(x) = x^2$ in terms of orthogonal set $\{\cos n\pi x\}_{n=0}^{\infty}$, $\{\sin n\pi x\}_{n=1}^{\infty}$, and $\{1, \cos n\pi x, \sin n\pi x\}_{n=1}^{\infty}$ in the interval $(-1, 1)$. Thus, there are three expansions. Can we expand the function in the interval $(-\pi, \pi)$ in terms of those functions? Why?

E4.7 Check whether the following Sturm-Liouville eigenvalue problems are regular or singular; and whether the eigenvalues are positive or not. If the problem is singular, where is the singularity?

(a) $((1+x^2)y')' + \lambda y = 0$, $y(0) = 0$, $y(3) = 0$.

(b) $xy'' + y' + \lambda y = 0$, $y(a) = 0$, $y(b) = 0$, $0 \leq a < b$.

(c) $xy'' + 2y' + \lambda y = 0$, $y(1) = 0$, $y'(2) = 0$. **Hint:** Multiply by x.

(d) $xy'' - y' + \lambda xy = 0$, $y(0) = 0$, $y'(5) = 0$. **Hint:** Divide by x^2.

(e) $((1-x^2)y')' - 2xy' + (1+\lambda x)y = 0$, $y(-1) = 0$, $y(1) = 0$.

Hint: First you need to re-write the problems as the standard Sturm-Liouville eigenvalue problems if possible.

E4.8 Find out the eigenvalues and eigenfunctions of the Sturm-Liouville eigenvalue problem. It is encouraged to use computers to plot first three eigenfunctions.

(a) $y'' + \lambda y = 0$, $y(0) = 0$, $y(4\pi) = 0$.

(b) $y'' + \lambda y = 0$, $y(0) = 0$, $y(\pi/4) = 0$.

(c) $y'' + \lambda y = 0$, $y'(0) = 0$, $y(4\pi) = 0$.

(d) $y'' + \lambda y = 0$, $y(0) = 0$, $y'(4\pi) = 0$.

(e) $y'' + \lambda y = 0$, $y(0) + y'(0) = 0$, $y(4\pi) = 0$.

(f) $y'' + \lambda y = 0$, $y(0) = 0$, $y(4\pi) + y'(4\pi) = 0$.

E4.9 Find out all eigenvalues and eigenfunctions of the Sturm-Liouville eigenvalue problem.

$$y'' + \lambda y = 0, \quad 0 < x < p,$$

$$y'(0) = 0, \quad y(p) = 0.$$

Plot first three eigenfunctions with $p = 1/2$, $p = 2$. Also answer the following questions:

(a) Can the eigenvalues of regular Sturm-Liouville eigenvalue problem be complex numbers?

(b) Are there finite or infinite number of distinct eigenvalues?

(c) If $y_m(x)$ and $y_n(x)$ are two eigenfunctions corresponding to two different eigenvalues λ_m and λ_n, what is the result of $\int_0^p y_m(x)y_n(x)dx$ if $m \neq n$? How about $\int_0^{p/2} y_m(x)y_n(x)dx$?

E4.10 Solve the 1D wave equation $\dfrac{\partial^2 u}{\partial t^2} = 4\dfrac{\partial^2 u}{\partial x^2}$ according to the following conditions:

(a) The Cauchy problem $-\infty < x < \infty$ with $u(x,0) = xe^{-x}$, $\dfrac{\partial u}{\partial t}(x,0) = e^{-x}$.

(b) The boundary value problem $u(0,t) = 0$, $u(2,t) = 0$, $0 < x < 2$ with $u(x,0) = \sin(4\pi x)$, $\dfrac{\partial u}{\partial t}(x,0) = \sin(4\pi x)$.

(c) The boundary value problem $u(0,t) = 0$, $u(2,t) = 0$, $0 < x < 2$ with $u(x,0) = x$, $\dfrac{\partial u}{\partial t}(x,0) = x^2$.

E4.11 Repeat the problem for the 1D heat equation $\dfrac{\partial u}{\partial t} = 4\dfrac{\partial^2 u}{\partial x^2}$ according to the following conditions:

(a) The boundary value problem $u(0,t) = 0$, $u(2,t) = 0$, $0 < x < 2$ with $u(x,0) = \sin(4\pi x)$.

(b) The boundary value problem $u(0,t) = 0$, $u(2,t) = 0$, $0 < x < 2$ with $u(x,0) = x$.

E4.12 Solve the 1D wave equation $\dfrac{\partial^2 u}{\partial t^2} = c^2\dfrac{\partial^2 u}{\partial x^2}$ with $\dfrac{\partial u}{\partial x}(0,t) = 0$, $u(L,t) = 0$ and the initial condition $u(x,0) = f(x)$, $\dfrac{\partial u}{\partial t}(x,0) = g(x)$. Find the solution when $c = 3$, $L = 2$, $g(x) = 1$, and $f(x) = \begin{cases} 1 & \text{if } 0 \leq x < 1, \\ 0 & \text{if } 1 \leq x \leq 2. \end{cases}$

E4.13 Solve the 1D heat equation $\dfrac{\partial u}{\partial t} = c^2\dfrac{\partial^2 u}{\partial x^2}$ with $u(0,t) = 0$, $\dfrac{\partial u}{\partial x}(L,t) = 0$ and the initial condition $u(x,0) = f(x)$.

(a) Let $u(x,t) = X(x)T(t)$, derive the equations for $X(x)$ and $T(t)$.

4.7. Exercises

(b) Solve the related Sturm-Liouville eigenvalue value problem for $X(x)$ first.

(c) Solve for $T(t)$ using the eigenvalues above.

(d) Find the series solution to the 1D heat equation.

(e) Find the solution when $c = 3$, $L = 2$, and $f(x) = \begin{cases} 1 & \text{if } 0 \leq x < 1, \\ 0 & \text{if } 1 \leq x \leq 2. \end{cases}$

Chapter 5

Various Fourier series, properties and convergence

We have seen that $\{\sin\frac{n\pi x}{L}\}$ and $\{\cos\frac{n\pi x}{L}\}$ play very important roles in the series of solution of partial differential equations of boundary value problems by the method of separation of variables. While these orthogonal functions are obtained from Sturm-Liouville eigenvalue problems, they should have reminded us of Fourier series in which $\{\sin nx\}$ and $\{\cos nx\}$ are used. Fourier series have wide applications in many areas of sciences and engineering particularly in electro-magnetics, signal processing, filter design, and fast computation using fast Fourier transforms (FFT). In this chapter, we will introduce various Fourier series, discuss the properties and convergence of those series. We will see three kinds of Fourier expansions of a function $f(x)$:

1 General Fourier expansions in $(-L, L)$

$$f(x) \sim \bar{a}_0 + \sum_{n=1}^{\infty}\left(a_n \cos\frac{n\pi x}{L} + b_n \sin\frac{n\pi x}{L}\right); \tag{5.1}$$

When $L = \pi$, we obtain the classical Fourier series. The reason to use $\bar{a}_0$ rather than a_0 can be seen later.

2 Half-range sine expansions in $(0, L)$

$$f(x) \sim \sum_{n=1}^{\infty} b_n \sin\frac{n\pi x}{L}; \tag{5.2}$$

3 Half-range cosine expansions $(0, L)$

$$f(x) \sim \bar{a}_0 + \sum_{n=1}^{\infty} a_n \cos\frac{n\pi x}{L}. \tag{5.3}$$

5.1 Period, piecewise continuous/smooth functions

We know that $\sin x, \cos x, \sin 2x, \cos 2x, \cdots$, are all period functions. The above three types of Fourier expansions all involve sine and cosine functions that are periodic. What is a period function? A function repeats itself in a fixed pattern.

Definition 5.1. *If there is a positive number T such at $f(x+T) = f(x)$ for any x, then $f(x)$ is called a period function with a period T.*

According to the definition, $f(x)$ should be defined in the entire space $(-\infty, \infty)$. Also, if $f(x) = f(x+T)$, then $f(x+2T) = f(x+T+T) = f(x+T) = f(x)$; and $2T$ is also a period of $f(x)$. To avoid the confusion, we only use the smallest such a $T > 0$, which is called the fundamental period, or simply the period, for short.

Example 5.1. *Find the period of $\sin x, \cos x, \tan x,$ and $\cot x$.*

The period of $\sin x, \cos x$ is 2π, while the period of $\tan x, \cot x$ is π.

Example 5.2. *Are the following functions periodic? If so, find the period of the functions,*
$$\cos \pi x, \ \sin x + \tan x, \ \sin x + \cos \frac{x}{2}, \ \sin x + x, \cos mx.$$

1. *Yes, $\cos \pi x = \cos \pi(x+T) = \cos(\pi x + \pi T)$; so the period is $T = 2$.*

2. *Yes, the sum of two periodic functions is still periodic; the period is the larger one, $T = 2\pi$.*

3. *Yes, $\sin x + \cos \frac{x}{2} = \sin(x+T) + \cos \frac{x+T}{2}$. Since the period of the second function is $T/2 = 2\pi$, we conclude that the period is $T = 4\pi$.*

4. *No, since x is not a periodic function.*

5. *Yes, from $\cos mx = \cos m(x+T) = \cos(mx + mT)$, we know that $mT = 2\pi$; thus the period is $T = \dfrac{2\pi}{m}$.*

Note that if $f(x) = C$, then it is a periodic function of any period including the zero. From the above example, we also know that the set $\{1, \cos \frac{nx}{L}\}_{n=1}^{\infty}$ has a common period $T = 2L\pi$, the largest period of all $\cos \frac{nx}{L}$ for all n's.

Example 5.3. *Let $f(x) = x - int(x) = x - [x]$, where $[x]$ is called a floor function, which means that $[x]$ is the greatest integer not larger than x, for example, $[1.5] = 1$,*

5.1. Period, piecewise continuous/smooth functions

$[0.5] = 0$, $[-1.5] = -2$, or the integer toward left. Then $f(x)$ is a period function with period $T = 1$. The period function can be expressed as

$$f(x) = x, \qquad \text{if } 0 \le x < 1, \\ = x - [x], \qquad \text{otherwise,} \tag{5.4}$$

or simply $f(x) = f(x+1)$ outside $[0, 1]$. Often it is enough to write down the function expression in one period and state that the function is periodic with the period specified. Figure 5.1 (a) is a plot of the integer (floor) function while Figure 5.1 (b) is a plot of the fraction part function that is a periodic with period $T = 1$.

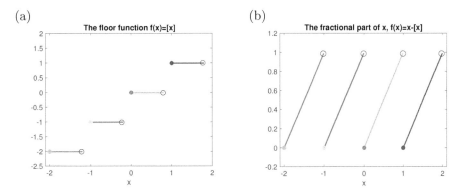

Figure 5.1. *Plots of some piecewise continuous functions. (a): the integer (floor) function that is not periodic. (b): the fractional part of x that can be expressed as $f(x) = x - int(x)$ which is periodic with period $T = 1$.*

Example 5.4. *The sawtooth function is defined by*

$$f(x) = \begin{cases} \frac{1}{2}(-\pi - x) & \text{if } -\pi \le x < 0, \\ \frac{1}{2}(\pi - x) & \text{if } 0 \le x \le \pi, \end{cases} \tag{5.5}$$

and $f(x) = f(x + 2\pi)$, see Figure 5.2 (b) for the function plot along its Fourier series and sum partial sums. Note that sometime it may be easier if we use the expression in the interval $(0, 2\pi)$ since it is one continuous piece as

$$f(x) = \frac{1}{2}(\pi - x), \qquad 0 \le x < 2\pi,$$

and $f(x) = f(x + 2\pi)$. Figure 5.2 is a plot of the sawtooth function whose Fourier series and some partial sums are plotted in Figure 5.6.

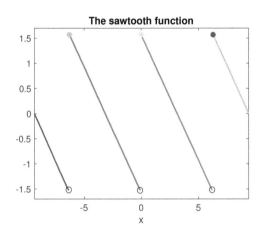

Figure 5.2. *Plots of the sawtooth function that is periodic with period $T = 2\pi$.*

Piecewise continuous/smooth functions

If a function $f(x)$ is continuous in $[a, b]$, then for any $x_0 \in [a, b]$, we have $\lim_{x \to x_0} f(x) = f(x_0)$. We call that $f(x)$ is in the continuous function space (set with operations) denoted as $f(x) \in C[a, b]$. It is obvious that functions: $\sin x, \cos x, x^3 + 1$, and their linear combinations are continuous functions in any interval $[a, b]$. The functions $f(x) = 1/x$ is discontinuous at $x = 0$, but is continuous on any interval that does not contain the origin. Note that $1/x$ is continuous on $(0, 1]$ but not $[0, 1]$. The function $\tan x$ is continuous on $[0, 1]$ but not on $[0, \frac{\pi}{2}]$ since the left limit $\lim_{x \to \frac{\pi}{2}-} f(x) = \infty$. From these examples, we should see the difference between '[' (included) and '(' (not included) in describing an interval.

If there are *finite* number of points $x_1, x_2, \cdots, x_N$ in $[a, b]$ at which a function is not continuous, but has finite left and right limits, that is

$$\lim_{x \to x_i-} f(x) = f(x_i-) \quad \text{and} \quad \lim_{x \to x_i+} f(x) = f(x_i+)$$

exist but $f(x_i-)$ may not be the same as $f(x_i+)$, then such a function is called a piecewise continuous function in (a, b), or precisely, a piecewise continuous and bounded function. Below is an example of a piecewise continuous and bounded

5.1. Period, piecewise continuous/smooth functions

function.

Example 5.5. *The Heaviside function* $H(x) = \begin{cases} 0 & \text{if } -\infty < x < 0, \\ 1 & \text{if } 0 \leq x < \infty, \end{cases}$ *is a piecewise continuous function, which is also called a* step *function.*

Figure 5.1, the floor function and the fractional part function, and Figure 5.2, the sawtooth function, are some examples pf piecewise continuous functions. Pay attention to when we use little hollow o's, and little filled o's.

If $f(x)$ is a continuous function on an interval (a, b), but $f'(x)$ is piecewise continuous on (a, b), then $f(x)$ is called a piecewise smooth function on (a, b). Below is an example.

Example 5.6. *The hat function* $h(x) = \begin{cases} 1 - |x| & \text{if } |x| \leq 1, \\ 0 & \text{otherwise,} \end{cases}$

is continuous but non-differentiable at $x = 0$ in the classical definition of derivatives. The derivative of the has function is

$$h'(x) = \begin{cases} 0 & \text{if } |x| > 1, \\ 1 & \text{if } -1 < x < 0, \\ -1 & \text{if } 0 < x < 1, \end{cases}$$

which is discontinuous at $x = -1$, $x = 0$, and $x = 1$. It is obvious that the hat function is a piecewise smooth function, while $h'(x)$ is a piecewise continuous function, which is also a step function. Figure 5.3 shows a plot of the hat function at the left, and the derivative of the hat function at the right.

Properties of period functions

The set of all period functions with the same period T form a linear space. That is, let $f(x)$ and $g(x)$ be two period functions of period T, then $w(x) = \alpha f(x) + \beta g(x)$ is also a period function of period T. Note again that a period function is defined in the entire space $-\infty < x < \infty$.

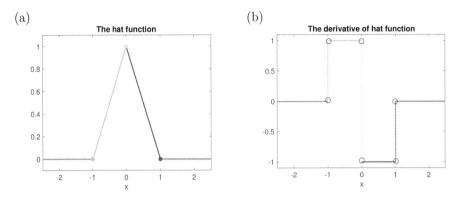

Figure 5.3. *Plot of the piecewise smooth hat function and its derivative. (a): the hat function; (b): the derivative of the hat function.*

Theorem 5.2. *Let $f(x)$ be a period function of period T and integrable, then*

$$\int_0^T f(x)dx = \int_a^{a+T} f(x)dx \tag{5.6}$$

for any real number a.

Proof: To prove the theorem, we just need to show that $\int_a^{a+T} f(x)dx$ is a constant function of a. Therefore, we define $F(a) = \int_a^{a+T} f(x)dx$ and take the derivative with respect to a to get,

$$\frac{dF(a)}{da} = f(a+T) - f(a) = 0.$$

Thus, $F(a)$ must be a constant, so $F(0) = F(a) = F(-T/2) = \cdots$, which leads to,

$$\int_0^T f(x)dx = \int_a^{a+T} f(x)dx = \int_{a-\frac{T}{2}}^{a+\frac{T}{2}} f(x)dx.$$

Often we prefer to use the period that

- $f(x)$ is a continuous piece;

- integration starts from the origin ($a = 0$);

- integration from a symmetric interval $(-\frac{T}{2}, \frac{T}{2})$.

5.2 The classical Fourier series expansion and partial sums

Let $f(x)$ be a periodic function of 2π and $f(x) \in L^2(-\pi, \pi)$. The classical Fourier series expansion of $f(x)$ is defined as

$$f(x) \sim \bar{a}_0 + \sum_{n=1}^{\infty} \left(a_n \cos nx + b_n \sin nx\right). \tag{5.7}$$

The coefficients $\bar{a}_0$, $\{a_n\}$ and $\{b_n\}$ are called the Fourier coefficients and can be computed from the following formulas,

$$\bar{a}_0 = \frac{1}{2\pi} \int_{-\pi}^{\pi} f(x) dx, \qquad a_n = \frac{1}{\pi} \int_{-\pi}^{\pi} f(x) \cos nx\, dx, \tag{5.8}$$

$$b_n = \frac{1}{\pi} \int_{-\pi}^{\pi} f(x) \sin nx\, dx, \qquad n = 1, 2, \cdots. \tag{5.9}$$

Note that $\bar{a}_0$ has different formula from other a_n's by a factor of 2. We can use the same formula if we use

$$f(x) \sim \frac{a_0}{2} + \sum_{n=1}^{\infty} \left(a_n \cos nx + b_n \sin nx\right). \tag{5.10}$$

We list a few applications of the Fourier series below:

- Express $f(x)$ in terms of simpler trigonometrical functions;

- Provide an approximation method for evaluating $f(x)$ using the partial sum defined as

$$S_N(x) = \bar{a}_0 + \sum_{n=1}^{N} \left(a_n \cos nx + b_n \sin nx\right), \tag{5.11}$$

as used in many computer packages for a given number N. We hope that $\lim_{N \to \infty} S_N(x) = f(x)$;

- Basis for several fast algorithms such as Fast Fourier Transform (FFT);

- Used for spectral (frequency) analysis, signal processing, filters, etc.

Note that if x is a time variable for some physical applications, we call that $f(x)$ is defined in the time domain, while $\{a_n\}_{n=0}^{\infty}$, $\{b_n\}_{n=1}^{\infty}$ are defined in the frequency domain.

Classical Fourier series of $f(x) \in L^2(a,b)$:

$$f(x) \sim \frac{a_0}{2} + \sum_{n=1}^{\infty}(a_n \cos nx + b_n \sin nx),$$

$$a_n = \frac{1}{\pi}\int_{-\pi}^{\pi} f(x) \cos nx \, dx, \quad b_n = \frac{1}{\pi}\int_{-\pi}^{\pi} f(x) \sin nx \, dx. \tag{5.12}$$

Example 5.7. *Find the classical Fourier series of $f(x) = x$.*

Solution: We may wonder $f(x) = x$ is not a periodic function and why it can have a Fourier expansion. In fact, we only use part of $f(x) = x$ in the interval $(-\pi, \pi)$ and disregard the rest (*truncation*). We then use the piece of $f(x) = x$ in the interval $(-\pi, \pi)$ to generate a periodic function (*extension*),

$$\tilde{f}(x) = \begin{cases} x & \text{if } |x| \leq \pi, \\ \tilde{f}(x+2\pi) & \text{otherwise,} \end{cases}$$

to get a periodic function that is identical to $f(x)$ in the interval $(-\pi, \pi)$, see Figure 5.4 for an illustration. The function $\tilde{f}(x)$ is piecewise continuous and bounded with discontinuities at $x = \pm 2n\pi, n = 1, 2, \cdots$. We use the formula to calculate the Fourier coefficients,

$$a_0 = \frac{1}{\pi}\int_{-\pi}^{\pi} f(x) dx = 0, \quad a_n = \frac{1}{\pi}\int_{-\pi}^{\pi} f(x) \cos nx \, dx = 0, \quad n = 1, 2, \cdots,$$

since $f(x)$ and $f(x)\cos nx$ are odd functions. Furthermore, we have

$$b_n = \frac{1}{\pi}\int_{-\pi}^{\pi} x \sin nx \, dx = \frac{2}{\pi}\int_0^{\pi} x \sin nx \, dx = -\frac{2}{n\pi} x \cos nx \Big|_0^{\pi} = (-1)^{n+1}\frac{2}{n}.$$

Thus, we get

$$x \sim \sum_{n=1}^{\infty}(-1)^{n+1}\frac{2}{n}\sin nx = 2\sin x - \sin 2x + \frac{2}{3}\sin 3x - \frac{1}{2}\sin 4x + \cdots.$$

From the formula (5.12), we can have the Fourier series for any function $f(x)$ on $(-\pi, \pi)$ literally as long as those integrations in the formula are finite. But the series may or may not converge, or converge to a value that is different from $f(x)$.

5.2. The classical Fourier series expansion and partial sums

To discuss the convergence of a Fourier series, we use the partial sums defined as before,

$$S_N(x) = f(x) \sim \frac{a_0}{2} + \sum_{n=1}^{N} (a_n \cos nx + b_n \sin nx). \qquad (5.13)$$

If the limit $\lim_{N \to \infty} S_N(x^*) = S(x^*)$ exists, then $S(x^*)$ is defined as the value of the series at x^*.

In Figure 5.4 (a), we plot the function $f(x) = x$ and a few partial sums of its Fourier series, $S_1(x)$, $S_5(x)$, $S_{55}(x)$ using the Maple. Figure 5.4 (b) is the function plot and the series plot in the interval $(-3\pi, 3\pi)$. From the figure, we can see that the partial sum $S_N(x)$ has the following properties:

1. converges to $f(x)$ in the interior of $(-\pi, \pi)$ as $N \to \infty$;

2. its value at $x = \pm\pi$ is not the left or right limit of $\tilde{f}(x)$, rather than its average, for example, at $x = \pi$,

$$\lim_{N \to \infty} S_N(\pi) = \frac{\tilde{f}(\pi-) + \tilde{f}(\pi+)}{2} = 0; \qquad (5.14)$$

3. $S_N(x)$ oscillates at the discontinuities $\pm 2n\pi$, $n = 1, 2, \cdots$. It is called the Gibb's phenomenon.

Note that the series itself is not oscillatory and it is identical to $f(x) = x$ in the interval $(-\pi, \pi)$, and it is zero at $x = \pm\pi$ which is the average of the left and right limits of the new period function $\tilde{f}(x)$, the *same* as the value of partial sums at $x = \pm\pi$, see Figure 5.4 (b).

Example 5.8. *Find the classical Fourier series of $f(x) = 10 \sin x + 5 \sin 6x + \frac{1}{2} \cos 30x$.*

Solution: The Fourier series of $f(x)$ is itself with $a_{30} = 0.5$, $b_1 = 10$, $b_6 = 5$. In Figure 5.5, we plot of the function and can see clearly the three frequencies and their strengths that agree with the function.

Example 5.9. *Find the Fourier series of the sawtooth function $f(x) = \frac{1}{2}(\pi - x)$, $0 \le x < 2\pi$, $f(x + 2\pi) = f(x)$.*

Note that $f(x)$ is an odd function in the interval of $(-\pi, \pi)$, thus $a_n = 0$, for $n = 0, 1, \cdots$. For the coefficients of b_n, it is easier to use one continuous piece in

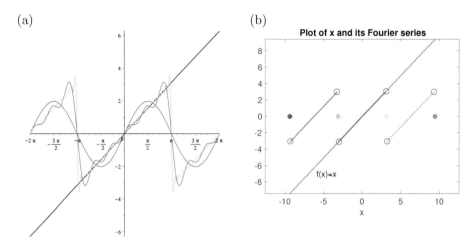

Figure 5.4. *(a): Plots of function $f(x) = x$ and some partial sums $S_1(x)$, $S_5(x)$, $S_{55}(x)$ of the Fourier series of the function. (b): Plots of the Fourier series of $\tilde{f}(x)$ whose values at $x = \pm 3\pi, \pm \pi$ are zero, and $f(x) = x$, the dotted line. Note that the two functions are identical in $(-\pi, \pi)$.*

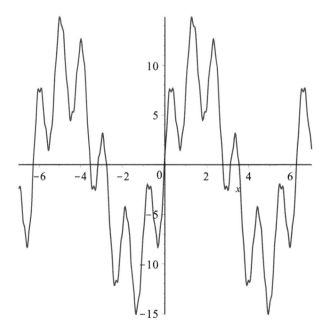

Figure 5.5. *Plot of $f(x) = 10\sin x + 5\sin 6x + \frac{1}{2}\cos 30x$. We can see clearly three different frequencies.*

5.2. The classical Fourier series expansion and partial sums

$(0, 2\pi)$ to calculate the coefficients b_n,

$$b_n = \frac{1}{\pi} \int_{-\pi}^{\pi} f(x) \sin nx \, dx = \frac{1}{\pi} \int_0^{2\pi} \frac{1}{2} (\pi - x) \sin nx \, dx$$

$$= \frac{1}{2\pi} \left(\int_0^{2\pi} \pi \sin nx \, dx + \frac{x \cos nx}{n} \bigg|_0^{2\pi} - \int_0^{2\pi} \frac{x \cos nx}{n} \, dx \right)$$

$$= \frac{1}{2\pi} \frac{2\pi}{n} = \frac{1}{n}.$$

Thus, the Fourier series is, see also Figure 5.2 (a),

$$\frac{1}{2}(x - \pi) \sim \sum_{n=1}^{\infty} \frac{\sin nx}{n} = \sin x + \frac{1}{2} \sin 2x + \frac{1}{3} \sin 3x + c \cdots + \frac{\sin nx}{n} + \cdots .$$

In Figure 5.2, we plot the function $f(x) = (\pi - x)$ of period 2π and a few partial sums. We observe that the partial sum converges to $f(x) = x - \pi$ except at those discontinuities at $x = 0$ and $x = \pm \pi$. Once again, we see the Gibb's oscillations of the partial sums around the discontinuities. It is also important to note that the series converges to $f(x)$ in the interval except at the discontinuities where the value of the series is the average of the left and right limits which is zero in this case. *There is no oscillations in the Fourier series though!*

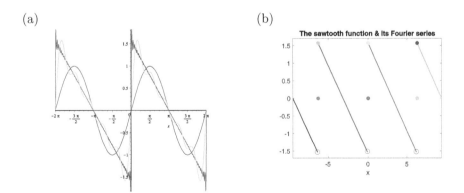

Figure 5.6. *(a): Plots of some partial sum of the Fourier series of the sawtooth function. (b): Plot the sawtooth function and its Fourier series. They are identical except at the discontinuities where the series is the average of the left and right limits.*

Example 5.10. *Find the Fourier series of the triangular wave.*

$$f(x) = \begin{cases} x + \pi & \text{if } -\pi \leq x < 0, \\ \pi - x & 0 \leq x \leq \pi, \end{cases} \tag{5.15}$$

and $f(x) = f(x + 2\pi)$. Note that we can rewrite the function as one piece as $f(x) = \pi - |x|$ in the interval $(-\pi, \pi)$, which is an even function in $(-\pi, \pi)$. The function is continuous in the interval and it is piecewise smooth.

Solutions: Note that $f(x)$ is an even function, we have

$$\bar{a}_0 = \frac{1}{2\pi} \int_{-\pi}^{\pi} f(x) dx = \frac{1}{\pi} \int_0^{\pi} (\pi - x) dx = -\frac{1}{\pi} \frac{(\pi - x)^2}{2} \bigg|_0^{\pi} = \frac{\pi}{2}$$

$$a_n = \frac{1}{\pi} \int_{-\pi}^{\pi} f(x) \cos nx\, dx = \frac{2}{\pi} \int_0^{\pi} (\pi - x) \cos nx\, dx$$

$$= \frac{2}{\pi} \left(\frac{(\pi - x) \sin nx}{n} \bigg|_0^{\pi} + \frac{1}{n} \int_0^{\pi} \sin nx\, dx \right)$$

$$= \frac{2}{\pi} \left(\frac{-\cos nx}{n^2} \bigg|_0^{\pi} \right) = \frac{2}{\pi} \left(\frac{1}{n^2} - \frac{(-1)^n}{n^2} \right)$$

$$= \frac{2}{\pi} \begin{cases} \dfrac{2}{n^2} & \text{if } n \text{ is odd}, \\ 0 & \text{if } n \text{ is even.} \end{cases}$$

We can use one simple notation $a_{2k+1} = \frac{4}{\pi(2k+1)^2}$ to cover both situations. Thus, we have,

$$f(x) = \frac{\pi}{2} + \sum_{n=0}^{\infty} \frac{4}{\pi(2n+1)^2} \cos \pi(2n+1)x.$$

Since $f(x)$ is continuous everywhere, we have the equality in the entire interval! In Figure 5.7, we plot the function of the triangular wave $f(x) = \pi - |x|$ of period 2π, and a few partial sums. We observe that the partial sum converges to $f(x) = \pi - |x|$ in the entire domain. We do not see the Gibb's oscillations but round-ups at the kinks, $x = 0$ and $x = \pm \pi$.

We can get some identities from the Fourier series. In this example, we have

$$f(0) = \pi = \frac{\pi}{2} + \sum_{n=0}^{\infty} \frac{4}{\pi(2n+1)^2}, \tag{5.16}$$

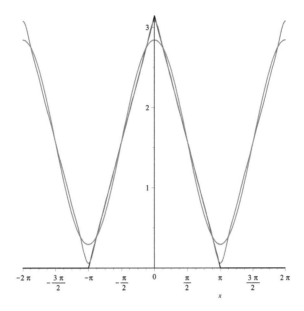

Figure 5.7. *Plot of the triangular wave and several partial sums. There is no Gibb's phenomenon but round-up at the kinks. The series is identical to the function of the triangular wave.*

which provide an alternative way in computing π if we multiply π from both sides and simplify to get

$$\frac{\pi^2}{2} = \sum_{n=0}^{\infty} \frac{4}{(2n+1)^2} \implies \frac{\pi^2}{8} = 1 + \frac{1}{3^2} + \frac{1}{2^2} + \frac{1}{7^2} + \cdots.$$

5.3 Fourier series of functions with arbitrary periods

Given a period function $f(x) \in L^2(-L, L)$ with $f(x+T) = f(x)$ and $T = 2L$, we can also have a Fourier series expansion of $f(x)$ in $(-L, L)$. To derive the Fourier series for a periodic function of $2L$, we use a linear transform to convert the interval $(-L, L)$ to $(-\pi, \pi)$, apply the Fourier expansion, and then transform back using the original variable.

Let $t = \alpha x$ and α is chosen such that when $x = -L$, $t = -\pi$, and when $x = L$, $t = \pi$. It is easy to get $\alpha = \frac{\pi}{L}$. Define also $f(x) = f(\frac{t}{\alpha}) = F(t)$. We can verify that $F(t)$ is a period function of 2π since

$$F(t + 2\pi) = f\left(\frac{t + 2\pi}{\alpha}\right) = f\left(\frac{t}{\alpha} + \frac{2\pi}{\alpha}\right) = f\left(\frac{t}{\alpha} + 2L\right) = f\left(\frac{t}{\alpha}\right) = F(t).$$

Thus, $F(t)$ has the Fourier series,

$$F(t) \sim \bar{a}_0 + \sum_{n=1}^{\infty} (a_n \cos nt + b_n \sin nt),$$

$$\bar{a}_0 = \frac{1}{2\pi} \int_{-\pi}^{\pi} F(t)dt, \qquad a_n = \frac{1}{\pi} \int_{-\pi}^{\pi} F(t) \cos nt\, dt,$$

$$b_n = \frac{1}{\pi} \int_{-\pi}^{\pi} F(t) \sin nt\, dt, \qquad n = 1, 2, \cdots.$$

By changing the variable again using $t = \frac{\pi}{L}$ in all the expressions above, we get

---

Fourier Series of $f(x)$ with an Arbitrary Period $2L$:

$$f(x) \sim \frac{a_0}{2} + \sum_{n=1}^{\infty} \left(a_n \cos \frac{n\pi x}{L} + b_n \sin \frac{n\pi x}{L} \right) \qquad (5.17)$$

$$a_n = \frac{1}{L} \int_{-L}^{L} f(x) \cos \frac{n\pi x}{L} dx, \qquad b_n = \frac{1}{L} \int_{-L}^{L} f(x) \sin \frac{n\pi x}{L} dx.$$

In the expression above we have a_n, $n = 0, 1, \cdots, \cdots$, and b_n, $n = 1, 2, \cdots$. Note that the above formula includes the classical Fourier series if we take $L = \pi$. Thus, it is enough just to remember this formula. Again the partial sum of the Fourier expansion in $(-L, L)$ is defined as

$$S_N(x) = \bar{a}_0 + \sum_{n=1}^{N} \left(a_n \cos \frac{n\pi x}{L} + b_n \sin \frac{n\pi x}{L} \right) \qquad (5.18)$$

for a positive integer $N > 0$.

Example 5.11. *Recall that the fractional part of x is a periodic function of period $T = 1$ and $p = 1$. The function can be written as $f(x) = x - \text{int}(x)$. We can find its Fourier series in $(-1, 1)$.*

Using the formula, we have

$$\bar{a}_0 = \frac{1}{2} \int_{-1}^{1} f(x)dx = \frac{1}{2} \left(\int_{-1}^{0} (x+1)dx + \int_{0}^{1} x dx \right) = \frac{1}{2},$$

$$a_n = \int_{-1}^{1} f(x) \cos \frac{n\pi x}{p} dx = \int_{-1}^{0} (x+1) \cos n\pi x\, dx + \int_{0}^{1} x \cos n\pi x\, dx = 0,$$

5.3. Fourier series of functions with arbitrary periods

$$b_n = \int_{-1}^{1} f(x) \sin \frac{n\pi x}{p} dx = \int_{-1}^{0} (x+1) \sin \frac{n\pi x}{p} dx + \int_{0}^{1} x \sin \frac{n\pi x}{p} dx$$

$$= \int_{-1}^{0} \sin \frac{n\pi x}{p} dx + 2\int_{0}^{1} x \sin \frac{n\pi x}{p} dx = -\frac{1}{n\pi}(1-(-1)^n) + \frac{1}{n\pi}(-1)^n$$

$$= \begin{cases} 0 & \text{if } n = 2k+1, \\ -\dfrac{2}{(n\pi)} & \text{if } n = 2k, \end{cases}$$

Thus, we obtain

$$f(x) = \frac{1}{2} - \sum_{k=1}^{\infty} \frac{1}{k\pi} \sin(2k\pi x).$$

In Figure 5.8, we plot the function $f(x) = x - int(x)$ and several partial sums of the Fourier series. The Fourier series converges to $f(x)$ in the interior. We observe the Gibb's oscillations at the discontinuity at $x = 0$, and the two end points $x = -1$ and $x = 1$.

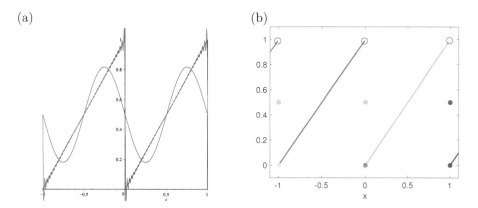

Figure 5.8. *(a): Plots of some partial sum of the Fourier series of the fraction part function, $f(x) = x - int(x)$. (b): Plot the function and its Fourier series. They are identical except at the discontinuity $x = 0$ and two end points $x = \pm 1$ where the series is the average of the left and right limits of the periodic function.*

Remark 5.1. *For the fractional part function example, the period is $T = 1$, we can also have the Fourier series in $(-\frac{1}{2}, \frac{1}{2})$ which will be different from that in $(-1, 1)$.*

Example 5.12. *Expand $f(x) = |x|$ in Fourier series in $(-p, p)$ for a parameter $p > 0$.*

Solution: Note that $f(x)$ is an even function, we have

$$\bar{a}_0 = \frac{1}{2p}\int_{-p}^{p} |x|dx = \frac{2}{2p}\int_0^p xdx = \frac{1}{p}\frac{x^2}{2}\bigg|_0^p = \frac{p}{2},$$

$$a_n = \frac{1}{p}\int_{-p}^{p} |x|\cos\frac{n\pi x}{p}dx = \frac{2}{2p}\int_0^p x\cos\frac{n\pi x}{p}dx$$

$$= -\frac{2p}{(n\pi)^2}(1-\cos n\pi) = \begin{cases} 0 & \text{if } n = 2k, \\ -\dfrac{4p}{(n\pi)^2} & \text{if } n = 2k+1, \end{cases}$$

$$b_n = \frac{1}{p}\int_{-p}^{p} |x|\sin\frac{n\pi x}{p}dx = 0,$$

since $f(x)$ and $f(x)\cos\frac{n\pi x}{p}$ are even functions, and $f(x)\sin\frac{n\pi x}{p}$ is an odd function. Thus, we obtain

$$|x| = \frac{p}{2} - \sum_{n=0}^{\infty} \frac{4p}{((2n+1)\pi)^2}\cos\frac{(2n+1)\pi x}{p}$$

$$= \frac{p}{2} - \frac{4p}{\pi^2}\left(\cos\frac{\pi x}{p} + \frac{1}{3^2}\cos\frac{3\pi x}{p} + \frac{1}{5^2}\cos\frac{5\pi x}{p} + \frac{1}{7^2}\cos\frac{7\pi x}{p} + \cdots\right).$$

In Figure 5.9, we take $p = 1$ and plot the function $f(x)$ and several partial sums of the Fourier series in the interval $(-2, 2)$. The Fourier series converges to $|x|$ only in the interval $[-1, 1]$ including the two end points. No Gibb's phenomenon is present for the partial sums since the function is piecewise smooth. But we do see that the kink of $|x|$ at $x = 0$ is smoothed by the partial sums, which are called *round-ups*.

When $p = \pi$, we get the classical Fourier series in $[-\pi, \pi]$,

$$|x| = \frac{\pi}{2} - \sum_{n=0}^{\infty} \frac{4}{(2n+1)\pi}\cos(2n+1)x$$

$$= \frac{\pi}{2} - \frac{4}{\pi}\left(\cos x + \frac{1}{3^2}\cos 3x + \frac{1}{5^2}\cos 5x + \frac{1}{7^2}\cos 7x + \cdots\right).$$

Remark 5.2. *In the expansion above, we expand the $2p$-function $f(x) = |x|$ and $f(x + 2p) = f(x)$ in terms of the Fourier series. The expansion is the same as*

5.3. Fourier series of functions with arbitrary periods

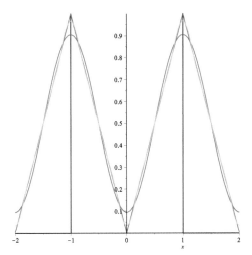

Figure 5.9. *Plot of the Fourier series and several partial sums of $|x|$ in the interval $(-2, 2)$ with $p = 1$. The Fourier series converges to $|x|$ only in the interval $[-1, 1]$.*

that for function $g(x) = |x|$ in the interval $(-p, p)$ but totally different outside the interval. There is no Gibb's phenomenon and the series is convergent to $|x|$ in $(-p, p)$. The process can be summarized as extension and expansion.

Example 5.13. *Expand $f(x) = \sin x$ in Fourier series in $(-p, p)$ for a parameter $p > 0$.*

Solution: If $p = \pi$, then the Fourier expansion of $\sin x$ is itself. Otherwise, we can expand $\sin x$ in terms of $\sin \frac{n\pi x}{p}$. Note that $a_n = 0$, $n = 0, 1, \cdots$, since $f(x)$ is an odd function. We just need to find b_n,

$$b_n = \frac{1}{p} \int_{-p}^{p} \sin x \sin \frac{n\pi x}{p} dx = \frac{2}{2p} \int_{0}^{p} \sin x \sin \frac{n\pi x}{p} dx$$

$$= \frac{2p \left(n\pi \sin p \cos n\pi - p \cos p \sin(n\pi) \right)}{p^2 - \pi^2 n^2}.$$

The integration is obtained by using the formula

$$\sin \alpha \sin \beta = -\frac{1}{2} \left(\cos(\alpha + \beta) - \cos(\alpha - \beta) \right)$$

or using the Maple command

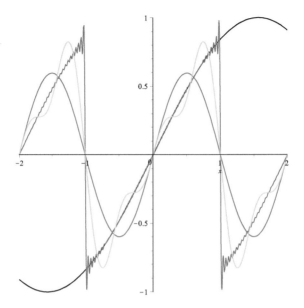

Figure 5.10. *Plots of several partial sums of the Fourier series of* $\sin x$ *in the interval* $(-2, 2)$ *with* $p = 1$. *The Fourier series converges to* $\sin x$ *only in the interval* $(-1, 1)$.

```
\int_0^p \sin x \sin \frac{n \pi x}{p} dx;
```

For the special case $p = 1$, we can get

$$\sin x = \sum_{n=1}^{\infty} (-1)^{n+1} \frac{2n\pi \sin 1 \sin(n\pi x)}{n^2 \pi^2 - 1},$$

which is valid only in the interval $(-1, 1)$, see Figure 5.10 for an illustration.

It is important to know the relations among the function itself, its Fourier series, and the partial sums. If the function $f(x)$ is continuous at a point x^*, then the series has the *same value* as that of $f(x)$. The partial sums are *approximations* to $f(x)$ and are different from $f(x)$ in general, that is, $f(x^*) \neq S_N(x^*)$. Nevertheless, the limit of the partial sum is $f(x^*)$, that is, $\lim_{N \to \infty} S_N(x^*) = f(x^*)$ if $f(x)$ is continuous at x^*. If $f(x)$ is discontinuous at a point x^*, then the value of the series is the average of the left and right limit, that is

$$\lim_{N \to \infty} S_N(x^*) = S(x^*) = \frac{\lim_{x \to x^*, x < x^*} f(x) + \lim_{x \to x^*, x > x^*} f(x)}{2} = \frac{f(x^*-) + f(x^*+)}{2}.$$

5.4. Half-range expansions

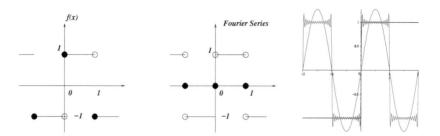

Figure 5.11. *Plots of $f(x)$, its Fourier series, and partial sums. Left: $f(x)$; Middle: Fourier series; Right: Two partial sums with $N = 2$, $N = 30$. Gibb's oscillations can been seen at the discontinuities if N is large enough.*

We use a step function below to illustrate the relations,

$$f(x) = \begin{cases} -1 & \text{if } -1 \leq x < 0, \\ 1 & 0 \leq x < 1. \end{cases} \tag{5.19}$$

In Figure 5.11, we plot the step function in the left; the Fourier series of the function in the middle; and two partial sums of the Fourier series in the right. We can see that $f(x)$ is piecewise continuous. The Fourier series is identical to $f(x)$ except at the discontinuities $x = -1$, $x = 0$, and $x = 1$. At these points, the Fourier series is the average of the left and right limit of the function, for example,

$$S(0) = \frac{f(0-) + f(0+)}{2} = \frac{-1 + 1}{2} = 0, \tag{5.20}$$

which is the same at $x = -1$ and $x = 1$. The partial sums are not the same as $f(x)$ but they will get closer to $f(x)$ as N increases except at those discontinuities at which the values are also the average of the left and right limit of the function. Note also that we will see the Gibb's oscillations around the discontinuities if N is large enough. Intuitively, the Fourier series tries to approximate both the left and right limit, which is impossible and causes the oscillations.

5.4 Half-range expansions

We have already seen that we can choose different expansions and seen some connections between Fourier series and orthogonal functions from the theory of Sturm-Liouville eigenvalue problems. With half-range expansion, we can also reduce some workload compared to a full range expansion, and impose some special properties of the expansions. The techniques once again is based some particular truncations and extensions.

Let $f(x)$ be a piecewise continuous function in $(0, L)$.[4] If we extend $f(x)$, $0 \leq x \leq L$ in the following way,

$$f_e(x) = \begin{cases} f(x) & \text{if } 0 \leq x \leq L, \\ f(-x) & \text{if } -L < x < 0, \end{cases} \tag{5.21}$$

which is called an *even extension* of $f(x)$, then we can have the Fourier series expansion of $f_e(x)$ in the interval $(-L, L)$. Since $f_e(x)$ is an even function, we have $b_n = 0$ and the expansion has cosine functions only

$$\bar{a}_0 = \frac{1}{2L} \int_{-L}^{L} f_e(x) dx = \frac{1}{L} \int_{0}^{L} f(x) dx, \tag{5.22}$$

$$a_n = \frac{1}{L} \int_{-L}^{L} f_e(x) \cos \frac{n\pi x}{L} dx = \frac{2}{L} \int_{0}^{L} f(x) \cos \frac{n\pi x}{L} dx. \tag{5.23}$$

Also in the interval, we have $f_e(x) = f(x)$, thus we obtain:

Half Range Cosine Series Expansion of $f(x)$ in $(0, L)$:

$$f(x) = \frac{a_0}{2} + \sum_{n=1}^{\infty} a_n \cos \frac{n\pi x}{L}, \quad a_n = \frac{2}{L} \int_{0}^{L} f(x) \cos \frac{n\pi x}{L} dx, \tag{5.24}$$

for $n = 0, 1, 2, \cdots$. The expansion is valid only in $(0, L)$.

Similarly, if we extend $f(x)$, $0 \leq x \leq L$ according to

$$f_o(x) = \begin{cases} f(x) & \text{if } 0 \leq x \leq L, \\ -f(-x) & \text{if } -L < x < 0, \end{cases} \tag{5.25}$$

which is called an *odd extension* of $f(x)$, then we can have the Fourier series expansion of $f_e(x)$ in the interval $(-L, L)$. Since $f_o(x)$ is an odd function, we have $a_n = 0$ and the expansion has sine functions only

$$b_n = \frac{2}{L} \int_{0}^{L} f(x) \sin \frac{n\pi x}{L} dx, \tag{5.26}$$

---

[4]In fact, the discussions are valid for any interval (a, b) ($b > a$). we can use a shift $s = x - a$ to change the domain from (a, b) in x to $(0, b - a)$ in terms of s.

5.4. Half-range expansions

$n = 1, 2, \cdots,$. Also in the interval, we have $f_o(x) = f(x)$, thus we have

---

Half Range Sine Series Expansion of $f(x)$ in $(0, L)$:

$$f(x) \sim \sum_{n=1}^{\infty} b_n \sin \frac{n\pi x}{L}, \quad b_n = \frac{2}{L} \int_0^L f(x) \sin \frac{n\pi x}{L} dx, \quad (5.27)$$

for $n = 1, 2, \cdots$. The expansion is valid only in $(0, L)$.

Example 5.14. *Expand $f(x) = x$ in both half-range cosine and sine series in $(0, 1)$. What is the relation of the expansion with the Fourier series in $(-1, 1)$.*

Solution: The function $f(x)$ is an odd function. Thus, the half-range sine series is the same as the Fourier series in $(-1, 1)$ for which the coefficients $a_n = 0$ and b_n can be calculated as (verified by Maple),

$$b_n = 2 \int_0^1 x \sin(n\pi x) dx = -2 \frac{\cos n\pi}{n\pi} = (-1)^n \frac{2}{n\pi}.$$

Thus, the sine expansion (and the Fourier expansion) of $f(x) = x$ in the interval $(0, 1)$ is

$$x = \sum_{n=1}^{\infty} (-1)^n \frac{1}{n\pi} \sin(n\pi x), \quad x \in (0, 1).$$

The series is convergent in the interior of $[0, 1)$ but is zero at $x = 1$ ($f(1) \neq S(1)$), see Figure 5.12 (b) for plots of the function, and several partial sums of the expansion. Note that the partial sums $S_N(x)$ have Gibb's oscillations near $x = 1$ if N is large enough.

For the cosine half-range expansion, the expansion is only valid in $[0, 1]$, we have

$$a_0 = \int_0^1 x\, dx = \frac{1}{2}, \quad a_n = 2\int_0^1 x\cos(n\pi x)dx = 2\frac{\cos n\pi - 1}{(n\pi)^2} = \frac{-4}{(2k-1)^2 \pi^2}.$$

Thus, the cosine expansion of $f(x) = x$ in the interval $(0, 1)$ can also be represented as

$$x = \frac{1}{2} - \frac{4}{\pi^2} \sum_{n=1}^{\infty} \frac{\cos(2n-1)\pi x}{(2n-1)^2}.$$

The series is convergent in the entire interval of $[0, 1]$, see Figure 5.12 (a) for plots of the function and several partial sums of the expansion. In this case, we have faster convergence of the partial sum of the cosine expansion compared with the sine expansion. Note that the partial sums $S_N(x)$ do not have Gibb's oscillations but round-ups near $x = 1$ if N is large enough.

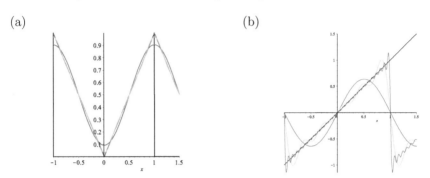

Figure 5.12. *Half-range cosine/sine Fourier expansions of $f(x) = x$ in $(0, 1)$ and plots of several partial sums. (a) Half cosine, the series is convergent in $[0, 1]$; (b) Half sine, the series is convergent in $[0, 1)$ but not to x at $x = 1$.*

Example 5.15. *Expand $f(x) = \cos x$ in both half-range cosine and sine series in $(0, \pi)$. What is the relation of the expansion with the Fourier series in $(-\pi, \pi)$. How about in $(0, 1)$?*

Solution: The function $f(x) = \cos x$ is an even function. Thus, the half-range cosine series is the same as the Fourier series in $(-\pi, \pi)$ or any 2π intervals, which is itself but it is different in $(-1, 1)$.

For the half-range sine expansion, we have (verified by Maple)

$$b_n = \frac{2}{\pi} \int_0^\pi \cos x \sin(nx) dx = \frac{2}{\pi} \frac{n(\cos n\pi + 1)}{n^2 - 1} = \frac{1}{\pi} \frac{8k}{(2k)^2 - 1},$$

where $n = 2k$ since for odd n's, $b_n = 0$. Thus, the sine expansion of $f(x) = \cos x$ in the interval $(0, \pi)$ is

$$\cos x = \sum_{n=1}^\infty \frac{1}{\pi} \frac{8n}{4n^2 - 1} \sin(2nx), \qquad x \in (0, \pi).$$

The series is convergent in the interior of $(0, \pi)$ but not to $\cos x$ at $x = 0$ and $x = \pi$, see Figure 5.13 (a) for plots of the function, and several partial sums of the expansion. Note that the partial sums $S_N(x)$ have Gibb's oscillations near $x = 0$ and $x = \pi$ if N is large enough.

We also plot the function and several partial sums of the cosine expansion of $\cos x$ in $(0, 1)$. In this case, the series is convergent in the entire interval $[0, 1]$ including two end points. The partial sums $S_N(x)$ have round-ups at $x = 1$ if N is large enough but no Gibb's oscillations.

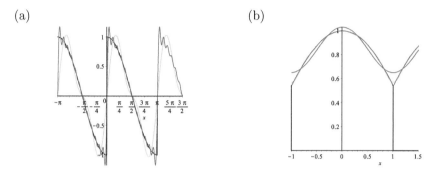

Figure 5.13. *Illustration of half-range sine/cosine Fourier expansions of $f(x) = \cos x$. (a): plots of the function, several partial sums of the half-range sine expansion on $(0, \pi)$. The series is convergent to $f(x) = \cos x$ in $(0, \pi)$ but not at two ends; (b): half-range cosine on $(0, 1)$. The series is convergent to $f(x) = \cos x$ on $[0, 1]$ including the two ends.*

5.5 Some theoretical results of various Fourier series

First of all, from the orthogonality of $\left\{\cos \frac{n\pi x}{L}\right\}_{n=0}^{\infty}$ and $\left\{\sin \frac{n\pi x}{L}\right\}_{n=1}^{\infty}$, we can easily prove the Parseval's identity.

Parseval's identity: If $f(x) \in L^2(-L, L)$ and
$$f(x) = \frac{a_0}{2} + \sum_{n=1}^{\infty}\left(a_n \cos \frac{n\pi x}{L} + b_n \sin \frac{n\pi x}{L}\right), \quad -L < x < L,$$
then the following Parseval's identity holds

Parseval's Identity
$$\frac{1}{2L}\int_{-L}^{L}|f(x)|^2 dx = \frac{a_0^2}{4} + \frac{1}{2}\sum_{n=1}^{\infty}\left(a_n^2 + b_n^2\right). \tag{5.28}$$

Note that the identity is not true for the integration in any interval but only for $(-L, L)$ for which the trigonometric functions on the right hand side is orthogonal!

Example 5.16. If $f(x) = \sum_{n=0}^{\infty} \frac{\cos nx}{2^n}$, find the value of $\int_{-\pi}^{\pi} f^2 dx$.

Solution: In this example, $L = \pi$, $a_0 = 2$, $a_n = \frac{1}{2^n}$, $b_n = 0$, thus we have

$$\frac{1}{2\pi}\int_{-L}^{L} |f(x)|^2 dx = 1 + \frac{1}{2}\sum_{n=1}^{\infty} \frac{1}{2^{2n}} = 1 + \frac{1}{2}\left(\frac{1/4}{1-1/4}\right) = \frac{7}{6}$$

$$\implies \int_{-\pi}^{\pi} |f(x)|^2 dx = \frac{7\pi}{3}.$$

From Parseval's identity, we can get some useful identities of series like the one above $\frac{1}{2}\sum_{n=1}^{\infty}\frac{1}{2^{2n}}$.

Now we discuss the calculus of Fourier series, which often deals with the limits, the differentiation, and integration of Fourier series. The tool is to use the partial sum of a series. We want to know whether the following is true.

$$\left(\lim_{x\to x_0};\ \frac{d}{dx};\ \int_{\alpha}^{\beta} dx\right) f(x) \stackrel{?}{=} \sum_{n=0}^{\infty} \left(\lim_{x\to x_0};\ \frac{d}{dx};\ \int_{\alpha}^{\beta} dx\right)\left(a_n \cos\frac{n\pi x}{L} + b_n \sin\frac{n\pi x}{L}\right).$$

The partial sum forms a sequence $\{S_0(x), S_1(x), S_2(x), \cdots, S_N(x), \cdots\}$ or $\{S_N(x)\}$. Note that $S_N(x)$ has two parameters, x and N. We will discuss two kinds of convergence, pointwise and uniform convergence in an interval. We will discuss more general sequence $f_n(x)$.

A pointwise convergence of $f_n(x)$ is defined for a fixed point x in an interval (a, b) such that $\lim_{n\to\infty} f_n(x) = f(x)$. For the partial sum $S_N(x)$, the pointwise convergence is the same as the convergence of the series.

Example 5.17. Are the following sequences convergent? (a), $f_n(x) = \frac{\sin nx}{n}$; (b), $g_n(x) = nxe^{-nx+1}$.

Solution: (a), $\lim_{n\to\infty} f_n(x) = \lim_{n\to\infty} \frac{\sin nx}{n} = 0$ for any x. (b), we can use the L'Hôspital's rule to get the limit, that is, $\lim_{n\to\infty} g_n(x) = \lim_{n\to\infty}\frac{nxe}{e^{nx}} = \lim_{n\to\infty}\frac{xe}{xe^{nx}} = 0$ for any $x \neq 0$, in which we differentiate n in the L'Hôspital's rule.

5.5. Some theoretical results of various Fourier series

In above examples, both $f_n(x)$ and $g_n(x)$ are convergent to zero in any interval. The function $f_n(x)$ gets smaller and smaller as n gets large, while there are always points x near zero such that $g_n(x) \sim 1$ no matter how large n can be. Such an $f_n(x)$ is also called uniformly convergent, while $g_n(x)$ is not uniformly convergent in the interval $(0, \pi)$. Note that $g_n(x)$ is uniformly convergent in any interval (a, b) if $a > 0$.

Definition 5.3. *Let $f_n(x)$ be a sequence defined in an interval $[a,b]$ and $f_n(x)$ has the pointwise convergence $\lim_{n \to \infty} f_n(x) = f(x)$ for any x in $[a,b]$. Given any number $\epsilon > 0$ (no matter how small it may be), if there is an integer N such that*

$$|f_n(x) - f(x)| < \epsilon \quad \text{for any } n > N \text{ and } x \text{ in } [a,b], \tag{5.29}$$

then $f_n(x)$ is called uniformly convergent to $f(x)$ in $[a,b]$.

In the previous example, given an $\epsilon > 0$, for $f_n(x) = \frac{\sin nx}{n}$, we have

$$|f_n(x)| = \left|\frac{\sin nx}{n}\right| \leq \left|\frac{1}{n}\right| < \epsilon,$$

as long as $n \geq \text{int}(1/\epsilon) + 1$. Thus, we can take $N = \text{int}(1/\epsilon) + 1$ after which we have $|f_n(x) - 0| < \epsilon$. However, for $g_n(x) = nxe^{-nx+1}$, no matter how large n is, we can find an $x = 1/n$ for which $g_n(x) = 1$ which can no be smaller than an arbitrarily small number ϵ. Thus, $g_n(x)$ is not uniformly convergent.

Definition 5.4. *For a given series $\sum_{n=0}^{\infty} u_n(x)$, if the partial sum $\{S_N(x)\}$ is uniformly convergent in an interval $[a,b]$, then the series is called uniformly convergent in the interval $[a,b]$.*

It is not easy to check whether a series is a uniformly convergent or not according to the definition. How do we know if a series is uniformly convergent without using the partial sum and the definition? The idea is to compare the series with a convergent series that does not have x in the series, which is always uniformly convergent. This is summarized in the Weierstrass M-test theorem.

Theorem 5.5. *Weierstrass M-test theorem. Given a series $\sum_{n=0}^{\infty} u_n(x)$ that satisfies* the following conditions

$$(i): \quad |u_n(x)| \leq M_n \quad \text{independent of } x \text{ in an interval } [a, b], \tag{5.30}$$

$$(ii): \quad \sum_{n=0}^{\infty} M_n < \infty \quad \text{the series that does not have } x \text{ converges}, \tag{5.31}$$

then the series is uniformly convergent in the interval $[a, b]$.

Example 5.18. *Are the following series uniformly convergent? Find intervals that the series are uniformly convergent.*

$$(a): \sum_{n=1}^{\infty} \frac{\sin nx}{n^2}, \quad (b): \sum_{n=1}^{\infty} \frac{\sin nx}{n}, \quad (c): \sum_{n=1}^{\infty} e^{-nx} \sin nx.$$

Solution: (a): We know that

$$\left|\frac{\sin nx}{n^2}\right| \leq \left|\frac{1}{n^2}\right| \quad \text{and the series} \quad \sum_{n=1}^{\infty} \frac{1}{n^2}$$

is convergent. Thus, the series is uniformly convergent. For (b), we can not use the Weierstrass M-test since the series $\sum_{n=1}^{\infty} \frac{1}{n}$ is divergent. So we do not have a conclusion about the uniform convergence since the theorem is a sufficient but not necessary. We will see the series cannot be uniformly convergent below. For (c), in any interval (a, b) where $a > 0$, we have

$$|e^{-nx} \sin nx| \leq e^{-nx} \leq e^{-na}, \quad \text{and} \quad \sum_{n=1}^{\infty} e^{-na}$$

is convergent. Thus, the series is uniformly convergent in (a, b) when $a > 0$. The series does not converge if $x < 0$, and it is convergent but not uniformly in any interval $(0, b)$ for $b > 0$.

Theorem 5.6. *If a series*

$$f(x) = \sum_{n=1}^{\infty} u_n(x)$$

is uniformly convergent in an interval (a, b), then the series, after we take a limit, or integrate, or differentiate, term by term, is still convergent in the interval (a, b), that is,

$$\left(\lim_{x \to x_0}; \frac{d}{dx}; \int_{\alpha}^{\beta} dx\right) f(x) = \sum_{n=0}^{\infty} \left(\lim_{x \to x_0}; \frac{d}{dx}; \int_{\alpha}^{\beta} dx\right) u_n(x).$$

5.6. Exercises

We may be able to use the theorem to further determine the uniform convergency of a series if the Weierstrass M-test fails. Reconsider the example (b) above for the series $\sum_{n=1}^{\infty} \frac{\sin nx}{n}$ for which the Weierstrass M-test fails. If the series was uniformly convergent, then we could take the derivative term by term to get a convergent series which is obviously untrue since the series $\sum_{n=1}^{\infty} \cos nx$ after the differentiation term by term diverges for almost any x. Thus, the series is not uniformly convergent. Another rule of thumb is that if the partial sums have Gibb's oscillations, then the series are not uniformly convergent. Therefore, the Fourier series for sawtooth function, the fractional part of x, and most half-range sine expansions are not uniformly convergent, while the Fourier series for triangular wave, and most half-range cosine expansions are uniformly convergent.

5.6 Exercises

E5.1 Find the period of the following functions. $f(x) = 5$, $f(x) = \cos x$, $f(x) = \cos(\pi x)$, $f(x) = \cos(px)$, $f(x) = \cos^2 x$, $f(x) = \cos x \sin x$, $f(x) = \cos(x/2) + 3\sin(2x)$, $f(x) = \tan(mx) + e^{\sin(2x)}$, and $\left\{\cos \frac{n\pi x}{L}\right\}_{n=0}^{\infty}$.

E5.2 Expand $f(x) = x$ and $g(x) = x^2$ as following series.

(a) (Half-range) sine series on $(0,2)$.

(b) (Half-range) cosine series on $(0,2)$.

(c) Fourier series on $(-2,2)$.

Check the pointwise and uniform convergence of all the series. Plot or sketch the series and the partial sums for large N.

E5.3 Given $\frac{\partial^2 u}{\partial t^2} - 2\frac{\partial^2 u}{\partial x^2} = 0$. Find the general solution, and also do the following.

(a) Solve the Cauchy problem $(-\infty < x < \infty)$ with $u(x,0) = |x|$, $|x| \leq 1$ and $u(x,0) = 0$ elsewhere; and $\frac{\partial u}{\partial t}(x,0) = 0$. Sketch the solution for $t = 5$.

(b) Solve the Cauchy problem $(-\infty < x < \infty)$ with $u(x,0) = e^{-x}\cos x$, $\frac{\partial u}{\partial t}(x,0) = \sin x$.

(c) Solve the boundary value problem $0 < x < 3$ with $u(0,t) = u(3,t) = 0$; $u(x,0) = \sin(6\pi x)$, $\frac{\partial u}{\partial t}(x,0) = \sin(24\pi x)$.

(d) Solve the boundary value problem $0 < x < 3$ with $u(0,t) = u(3,t) = 0$; $u(x,0) = x+1$, $\dfrac{\partial u}{\partial t}(x,0) = \cos x$.

E5.4 Let $f(x) = x$, $-\pi \leq x < \pi$ be a 2π-periodic function.

(a) Sketch the function in the interval $[-3\pi, 3\pi]$.

(b) Find the Fourier Series Expansion for $f(x)$. Sketch the partial sums $S_N(x)$ for large N, and the Fourier series.

(c) Are the formulas

$$b_n = \frac{1}{\pi}\int_{-\pi}^{\pi} f(x)\sin(nx)\,dx, \quad b_n = \frac{1}{\pi}\int_{0}^{2\pi} f(x)\sin(nx)\,dx$$

the same? Why? Evaluate the second integration.

(d) Find the common period of $1, \cos(nx), \sin(nx)$, $n = 1, 2, \cdots$.

E5.5 Expand $f(x) = \cos(2x)$ according to the following. **Check** whether the series are uniformly convergent or not; sketch the partial sums $S_N(x)$ with large N and the series (mark Gibb's oscillations or round-ups if applies); and write down the Parseval's identity.

(a) The classical Fourier series in $(-\pi, \pi)$.

(b) The Fourier series in $(-1, 1)$.

(c) Half-range sine series in $(0, \pi)$.

(d) Half-range cosine series in $(0, 1)$.

(e) Half-range cosine series in $(0, \pi)$.

Note: Pay attention to normal modes. It is okay to use Maple in some cases. If you do by hand, you need to evaluate the integrals.

E5.6 Write down the Parseval's identity corresponding to the classical Fourier series of $f(x) = x^2$.

E5.7 Find the Fourier series of the following $f(x)$ in $(-p, p)$, sketch $f(x)$, the partial sums $S_N(x)$ with large N, and the Fourier series in $(-p, p)$.

(a) $f(x) = -1$, $-p \leq x < 0$; $f(x) = 1$, $0 \leq x < p$.

(b) $f(x) = a\left(1 - \left(\frac{x}{p}\right)^2\right)$, $-p \leq x < p$.

(c) $f(x) = \begin{cases} c & \text{if } |x| \leq d, \\ 0 & d < |x| < p, \end{cases}$ where $0 < d < p$.

5.6. Exercises

E5.8 Find the half-range **sine and cosine** expansions of function $f(x)$ in $(0,\ p)$, sketch $f(x)$, the partial sums $S_N(x)$ with large N, and the Fourier series in $(0,\ p)$

(a) $f(x) = 1$, $p = 1$.

(b) $f(x) = 1$, $p = 3\pi$.

(c) $p = 2$, $f(x) = \begin{cases} 0 & \text{if } 0 < x \leq 1, \\ x - 1 & 1 < x < 2. \end{cases}$

(d) $f(x) = \sin(\pi x)\cos(\pi x)$, $p = 2\pi$.

E5.9 (a). Use the Parseval's identity and the Fourier series expansion of $f(x) = x/2$ in $(-\pi, \pi)$,

$$\frac{x}{2} = \sum_{n=1}^{\infty} \frac{(-1)^{n+1}}{n} \sin nx$$

to obtain $\sum_{n=1}^{\infty} \dfrac{1}{n^2} = \dfrac{\pi^2}{6}$.

(b). From (a) to obtain that $\sum_{k=1}^{\infty} \dfrac{1}{(2k)^2} = \dfrac{\pi^2}{24}$.

(c). Combine (a) and (b) to derive the identity $\sum_{k=0}^{\infty} \dfrac{1}{(2k+1)^2} = \dfrac{\pi^2}{8}$.

E5.10 Compute $\int_{-\pi}^{\pi} f^2(x)dx$ using the Parseval's identity.

(a) : $f(x) = \sum_{n=1}^{\infty} \dfrac{\cos nx}{n^2}$;   (b) : $f(x) = 1 + \sum_{n=1}^{\infty} \left(\dfrac{\cos nx}{3^n} + \dfrac{\sin nx}{n}\right)$.

Hint: Use the geometric series and the table of Zeta functions, see

https://en.wikipedia.org/wiki/Particular_values_of_the_Riemann_zeta_function.

E5.11 Use the Weierstrass M-test to judge whether the following series are uniformly convergent or not.

(a) $\sum_{n=1}^{\infty} \left(\dfrac{\cos nx}{n^2} + \dfrac{\sin nx}{n^3}\right)$.

(b) $\sum_{n=1}^{\infty} \dfrac{x^n}{n!}$, $|x| \leq 10$.

(c) $\sum_{n=1}^{\infty} \dfrac{\cos(x/n)}{n}$ for all x.

Which series are continuous, differentiable, and integrable, on which intervals?

E5.12 Group work: Design a wave filter that can only certain frequencies to pass, say $[a, b]$. Use this example to check

$$W(x) = 2.5\sin x + 4\cos(3x) + 10\cos(20(x+\pi)) + 0.02\sin(100*x) + \epsilon rand(x).$$

In the above, $rand(x)$ is a random number generator. In the above function, which frequency is dominant? If $\epsilon = 0$, what is the Fourier series of $W(x)$? Let $\epsilon = 10^{-5}$, find the Fourier series of $W(x)$ then carry out the filtering.

Chapter 6

Series solutions of PDEs of boundary value problems

In this chapter, we continue to discuss the method of separation of variables for various boundary value problems of partial differential equation. We will see more relations of the solution with the Sturm-Liouville eigenvalue problems, orthogonal expansions, and various Fourier series.

6.1 One-dimensional wave equations

Recall one-dimensional wave equations,

$$\frac{\partial^2 u}{\partial t^2} = c^2 \frac{\partial^2 u}{\partial x^2}, \qquad (6.1)$$

where $c > 0$ is called a wave number in physics. We have already known how to solve the problem for various situations.

- The general solution is $u(x,t) = F(x-ct) + G(x+ct)$, where $F(x)$ and $G(x)$ are two arbitrarily differentiable functions, that is, no conditions are attached to the partial differential equation.

- Solution to a Cauchy problem, that is, given $u(x,0) = f(x)$; $\frac{\partial u}{\partial t}(x,0) = g(x)$, $\infty < x < \infty$, the solution is given by the D'Alembert's formula

$$u(x,t) = \frac{1}{2}\left(f(x-ct) + f(x+ct)\right) + \frac{1}{2c}\int_{x-ct}^{x+ct} g(s)ds.$$

- Solution to some boundary value problems with normal modes initial conditions, that is, given an interval $0 < x < L$, the normal modes solution for some special $u(x,0) = f(x)$ and $u_t(x,0) = g(x)$, which can be expressed as

$$f(x) = \sum_{n=1}^{N} a_n \sin \frac{n\pi x}{L}, \quad g(x) = \sum_{n=1}^{N} b_n \sin \frac{n\pi x}{L}.$$

The solution is

$$u(x,t) = \sum_{n=1}^{N} \left(a_n \sin \frac{n\pi x}{L} \cos \frac{n\pi c t}{L} + \frac{b_n}{n\pi c} \sin \frac{n\pi x}{L} \sin \frac{n\pi c t}{L} \right).$$

- General initial conditions, $u(x,0) = f(x)$, $\frac{\partial u}{\partial t}(x,0) = g(x)$, $0 < x < L$. The solution is

$$u(x,t) = \sum_{n=1}^{\infty} \sin \frac{n\pi x}{L} \left(b_n \cos \frac{c n \pi t}{L} + b_n^* \sin \frac{c n \pi t}{L} \right),$$

where the coefficients are determined by

$$b_n = \frac{2}{L} \int_0^L f(x) \sin \frac{n\pi x}{L} dx,$$

$$b_n^* = \frac{2}{c n \pi} \int_0^L g(x) \sin \frac{n\pi x}{L} dx.$$

Note that the coefficients b_n's are obtained from the half-range sine expansion of $f(x)$ and b_n^*'s are obtained from the half-range sine expansion of $g(x)$ by a constant that depends on n.

Now we discuss how to solve more general one-dimensional wave equations.

Example 6.1. *An example with non-homogeneous boundary condition.*

$$\frac{\partial^2 u}{\partial t^2} = c^2 \frac{\partial^2 u}{\partial x^2}, \quad 0 < x < L,$$

$$u(0,t) = u_0, \quad u(L,t) = u_0,$$

$$u(x,0) = f(x), \quad \frac{\partial u}{\partial t}(x,0) = g(x), \quad 0 < x < L.$$

In this case, we can use the transformation

$$v(x,t) = u(x,t) - u_0$$

to get the homogenous BC for $v(x,t)$

$$\frac{\partial^2 v}{\partial t^2} = c^2 \frac{\partial^2 v}{\partial x^2}, \quad 0 < x < L,$$

$$v(0,t) = 0, \quad v(L,t) = 0,$$

$$v(x,0) = f(x) - u_0, \quad \frac{\partial v}{\partial t}(x,0) = g(x), \quad 0 < x < L.$$

The solution then will be

$$u(x,t) = u_0 + \sum_{n=1}^{\infty} \sin\frac{n\pi x}{L}\left(b_n \cos\frac{cn\pi t}{L} + b_n^* \sin\frac{cn\pi t}{L}\right)$$

where the coefficients are determined by

$$b_n = \frac{2}{L}\int_0^L f(x)\sin\frac{n\pi x}{L}dx,$$

$$b_n^* = \frac{2}{cn\pi}\int_0^L g(x)\sin\frac{n\pi x}{L}dx.$$

Challenge: How about different boundary condition $\frac{\partial u}{\partial x}(0,t) = 0$ and $u(L,t) = 0$. What are the normal mode solutions?

6.2 Series solution of 1D wave equations with derivative boundary conditions

An example with a Neumann boundary condition is given below,

$$\frac{\partial^2 u}{\partial t^2} = c^2 \frac{\partial^2 u}{\partial x^2}, \quad 0 < x < L,$$

$$u(0,t) = 0, \quad \frac{\partial u}{\partial x}(L,t) = 0,$$

$$u(x,0) = f(x), \quad \frac{\partial u}{\partial t}(x,0) = g(x), \quad 0 < x < L,$$

We will get a different Sturm-Liouville eigenvalue problem and a different expansion.

Step 1: Let $u(x,t) = T(t)X(x)$ and plug its partial derivatives into the original PDE so that we can separate variables. The homogeneous boundary conditions require $X(0) = X'(L) = 0$. Differentiating $u(x,t) = T(t)X(x)$ with respect to t and x, respectively, we get

$$\frac{\partial u}{\partial t} = T'(t)X(x), \quad \frac{\partial^2 u}{\partial t^2} = T''(t)X(x);$$

$$\frac{\partial u}{\partial x} = T(t)X'(x), \quad \frac{\partial^2 u}{\partial x^2} = T(t)X''(x).$$

The wave equation can be re-written as

$$T''(t)X(x) = c^2 T(t)X''(x) \quad \Longrightarrow \quad \frac{T''(t)}{c^2 T(t)} = \frac{X''(x)}{X(x)} = -\lambda. \tag{6.2}$$

This is because in the last equality, the left hand side is a function of t while the right hand side is a function of x, which is possible only both of them are a constant independent of t and x. We can get eigenvalues either for $X(x)$ or $T(t)$. Since we know the boundary condition for $X(x)$, naturally we should solve

$$\frac{X''(x)}{X(x)} = -\lambda \quad \text{or} \quad X''(x) + \lambda X(x) = 0, \quad X(0) = 0, \quad X'(L) = 0 \quad (6.3)$$

first.

Step 2: Solve the eigenvalue problem. From the Sturm-Liouville eigenvalue theory, we know that $\lambda > 0$ since the $q(x)$ in the S-L eigenvalue problem is zero. Thus the solution of $x(x)$ is,

$$X''(x) = C_1 \cos\sqrt{\lambda}x + C_2 \sin\sqrt{\lambda}x.$$

From the boundary condition $X(0) = 0$, we get $C_1 = 0$. From the boundary condition $X'(L) = 0$, we get

$$C_2 \cos\sqrt{\lambda}L = 0, \quad \Longrightarrow \quad \sqrt{\lambda}L = \frac{\pi}{2} + n\pi, \quad n = 0, 1, 2, \cdots,$$

since $C_2 \neq 0$. Note that, different from before, we should include $n = 0$. The eigenvalues and their corresponding eigenfunctions are

$$\lambda_n = \left(\frac{(2n+1)\pi}{2L}\right)^2, \quad X_n(x) = \sin\frac{(2n+1)\pi x}{2L}, \quad n = 0, 1, 2, \cdots.$$

Now we solve for $T(t)$ using

$$T''(t) + c^2 \lambda_n T(t) = 0. \quad (6.4)$$

The solution (not an eigenvalue problem anymore since we have already known λ_n) of $T(t)$ is,

$$T_n(t) = b_n \cos\frac{(2n+1)\pi ct}{2L} + b_n^* \sin\frac{(2n+1)\pi ct}{2L}.$$

Putting $X_n(x)$ and $T_n(t)$ together, we get one normal mode solution with each n

$$u_n(x,t) = \sin\frac{(2n+1)\pi x}{2L} \left(b_n \cos\frac{(2n+1)\pi ct}{2L} + b_n^* \sin\frac{(2n+1)\pi ct}{2L}\right), \quad (6.5)$$

which satisfy the PDE, the boundary conditions, but not the initial conditions.

Step 3: Put all the normal mode solutions together to get the series solution using the superposition. The coefficients are obtained from the orthogonal expansion of the initial conditions. The solution to the IVP-BVP of the 1D wave equation with a derivative boundary condition can be written as

$$u(x,t) = \sum_{n=0}^{\infty} \sin\frac{(2n+1)\pi x}{2L} \left(b_n \cos\frac{(2n+1)\pi ct}{2L} + b_n^* \sin\frac{(2n+1)\pi ct}{2L}\right) \quad (6.6)$$

6.2. Series solution of 1D wave equations with derivative boundary conditions

which satisfies the PDE and the boundary conditions. The coefficients of b_n and b_n^* are determined from the initial conditions $u(x,0) = f(x)$ and $u_t(x,0) = g(x)$,

$$u(x,0) = \sum_{n=0}^{\infty} b_n \sin \frac{(2n+1)\pi x}{2L} \implies b_n = \frac{2}{L}\int_0^L f(x) \sin \frac{(2n+1)\pi x}{2L} dx$$

$$\frac{\partial u}{\partial t}(x,0) = \sum_{n=0}^{\infty} \sin \frac{(2n+1)\pi x}{2L} \left(-b_n \frac{(2n+1)\pi c}{2L} \sin \frac{(2n+1)\pi ct}{2L} \right.$$

$$\left. + b_n^* \frac{(2n+1)\pi c}{2L} \cos \frac{(2n+1)\pi ct}{2L} \right),$$

$$\frac{\partial u}{\partial t}(x,0) = \sum_{n=0}^{\infty} \sin \frac{(2n+1)\pi x}{2L} b_n^* \frac{(2n+1)\pi c}{2L} \implies$$

$$b_n^* = \frac{\frac{2L}{(2n+1)\pi c} \int_0^L g(x) \sin \frac{(2n+1)\pi x}{2L} dx}{\int_0^L \left(\sin \frac{(2n+1)\pi x}{2L} \right)^2 dx}$$

$$= \frac{4}{(2n+1)\pi c} \int_0^L g(x) \sin \frac{(2n+1)\pi x}{2L} dx.$$

6.2.1 Summary of series solutions of 1D wave equations with homogeneous linear BC's

From above discussions, we can summarize the series solutions to 1D wave equations with different boundary conditions below.

Series solutions of 1D wave equations

$$\frac{\partial^2 u}{\partial t^2} = c^2 \frac{\partial^2 u}{\partial x^2}, \quad u(x,0) = f(x); \quad \frac{\partial u}{\partial t}(x,0) = g(x), \quad 0 < x < L,$$

with homogeneous Dirichlet, Neumann, and Robin boundary conditions have the following uniform form.

$$u(x,t) = \sum_{n=1 \text{ or } 0}^{\infty} X_n\left(\frac{\alpha_n \pi x}{L}\right) \left(b_n \cos \frac{\alpha_n \pi ct}{L} + b_n^* \sin \frac{\alpha_n \pi ct}{L} \right) \quad (6.7)$$

where

$$b_n = \frac{2}{L} \int_0^L f(x) X_n\left(\frac{\alpha_n \pi x}{L}\right) dx, \quad b_n^* = \frac{2}{\alpha_n \pi c} \int_0^L g(x) X_n\left(\frac{\alpha_n \pi x}{L}\right) dx. \quad (6.8)$$

Dirichlet-Dirichlet: $u(0,t) = u(L,t) = 0$. We have $X_n\left(\frac{\alpha_n \pi x}{L}\right) = \sin\frac{n\pi x}{L}$.

Dirichlet-Neumann:

$u(0,t) = 0, \ \frac{\partial u}{\partial x}(L,t) = 0.$ We have $X_n\left(\frac{\alpha_n \pi x}{L}\right) = \sin\frac{(\frac{1}{2}+n)\pi x}{L}$.

Neumann-Dirichlet:

$\frac{\partial u}{\partial x}(0,t) = 0, \ u(L,t) = 0.$ We have $X_n\left(\frac{\alpha_n \pi x}{L}\right) = \cos\frac{(\frac{1}{2}+n)\pi x}{L}$.

Neumann-Neumann: Any constants are solutions so the solution is not unique.

Dirichlet-Robin: $u(0,t) = 0, \ u(L,t) + \frac{\partial u}{\partial x}(L,t) = 0$. We have $X_n\left(\frac{\alpha_n \pi x}{L}\right) = \sin\frac{\alpha_n \pi x}{L}$, where we do not have analytic expressions for α_n.

For Robin-Dirichlet, or Neumann-Robin, or other linear boundary conditions, we generally do not have an analytic form for the eigenvalues and eigenfunctions. Thus, there are no analytic series solutions available.

6.2.2 Series solution of 1D wave equations of BVPs with a lower order term

Consider an example of a 1D wave equation with a lower order term of the following,

$$\frac{\partial^2 u}{\partial t^2} + a^2 u = c^2 \frac{\partial^2 u}{\partial x^2}, \quad 0 < x < L,$$

$$u(0,t) = 0, \quad u(L,t) = 0,$$

$$u(x,0) = f(x), \quad \frac{\partial u}{\partial t}(x,0) = g(x).$$

Solution: The method of separation variables with $u(x,t) = T(t)X(x)$ will lead to

$$\frac{T''(t) + a^2 T(t)}{c^2 T(t)} = \frac{X''(x)}{X(x)} = -\lambda. \tag{6.9}$$

We still have $\lambda_n = \left(\frac{n\pi}{L}\right)^2$ and $X_n(x) = \sin\frac{n\pi x}{L}$. But the solution of $T_n(t)$ will be different.

$$T_n(t) = b_n \cos\left(\sqrt{a^2 + \frac{c^2 n^2 \pi^2}{L^2}}\, t\right) + b_n^* \sin\left(\sqrt{a^2 + \frac{c^2 n^2 \pi^2}{L^2}}\, t\right). \quad (6.10)$$

The solution then is

$$u(x,t) = \sum_{n=1}^{\infty} \sin\frac{n\pi x}{L}\left\{ b_n \cos\left(\sqrt{a^2 + \frac{c^2 n^2 \pi^2}{L^2}}\, t\right) + b_n^* \sin\left(\sqrt{a^2 + \frac{c^2 n^2 \pi^2}{L^2}}\, t\right)\right\},$$

with b_n being the coefficient of the half-range sine expansion of $f(x)$

$$b_n = \frac{2}{L}\int_0^L f(x) \sin\frac{n\pi x}{L} dx, \quad (6.11)$$

and

$$b_n^* = \frac{2}{L\alpha_n}\int_0^L g(x) \sin\frac{n\pi x}{L} dx, \quad \text{where} \quad \alpha_n = \sqrt{a^2 + \frac{c^2 n^2 \pi^2}{L^2}}. \quad (6.12)$$

Challenge for a group work: How about the modified PDE below

$$\frac{\partial^2 u}{\partial t^2} + au = c^2 \frac{\partial^2 u}{\partial x^2}, \quad 0 < x < L,$$

assuming a is an arbitrary constant with the same boundary and initial conditions above? The solution for $T(t)$ may have two parts, the exponential functions and trigonometric functions.

6.3 Series solution to 1D heat equations with various BC's

A one-dimensional (1D) heat equation

$$\frac{\partial u}{\partial t} = c^2 \frac{\partial^2 u}{\partial x^2}, \quad (6.13)$$

is a good mathematical model of the temperature distribution in a rod in which c^2 is called the heat conductivity. The above partial differential equation is a second order, constant coefficients, linear, and homogeneous one. The PDE is classified as a *parabolic PDE*. We can check that

$$u(x,t) = \frac{1}{\sqrt{4\pi c^2 t}}\, e^{-\frac{x^2}{4c^2 t}} \quad (6.14)$$

is a solution to the PDE. It is called a *fundamental solution to the heat equation* in the place of general solutions for advection and wave equations. The fundamental solution of the heat equation corresponds to an instant heat source at $(0,0)$. The heat will be felt anywhere anytime instantly. Thus, the solution to a heat equation is called global one meaning that a change at any place at any time will affect the solution everywhere and anytime after. This is in contrast to solutions to advection and wave equations.

The solution to the Cauchy problem

$$\frac{\partial u}{\partial t} = c^2 \frac{\partial^2 u}{\partial x^2}, \quad -\infty < x < \infty, \quad (6.15)$$
$$u(x,0) = f(x),$$

is the convolution of $f(x)$ and the fundamental solution,

$$u(x,t) = \int_{-\infty}^{\infty} \frac{f(\xi)}{\sqrt{4c^2\pi t}} e^{-\frac{(x-\xi)^2}{4c^2 t}} d\xi. \quad (6.16)$$

Note that there is only one initial condition since the PDE involves only the first order derivative of the solution with time t.

Now we discuss the initial and boundary value problems for one-dimensional heat equation with various boundary conditions. Note that for homogeneous Dirichlet boundary condition $u(0,t) = 0$ and $u(L,t) = 0$, the derivation and the formula of the series solution have been given in Section 4.6. We first review one example here.

Example 6.2. *We can solve some 1D heat equations of boundary value problems with normal mode initial conditions as in the example below.*

$$\frac{\partial u}{\partial t} = 2\frac{\partial^2 u}{\partial x^2}, \quad 0 < x < 3,$$

$$u(0,t) = 0, \quad u(3,t) = 0,$$

$$u(x,0) = 5\sin(4\pi x) - 3\sin(8\pi x) + 2\sin(10\pi x).$$

Solution: It is clear that we can use the normal mode solutions for this problem with $L = 3$, $c^2 = 2$. The initial condition can be written as

$$u(x,0) = B_1 \sin\frac{m_1 \pi x}{3} + B_2 \sin\frac{m_2 \pi x}{3} + B_3 \sin\frac{m_3 \pi x}{3}$$
$$= 5\sin(4\pi x) - 3\sin(8\pi x) + 2\sin(10\pi x),$$

6.3. Series solution to 1D heat equations with various BC's

which is possible if and only if $B_1 = 5$, $m_1 = 12$, $B_2 = -3$, $m_2 = 24$, and $B_3 = 2$, $m_3 = 30$. From the formula

$$u(x,t) = \sum_{n=1}^{\infty} b_n \sin \frac{n\pi x}{L} e^{-c^2(\frac{n\pi}{L})^2 t},$$

we get the solution

$$u(x,t) = 5e^{-32\pi^2 t}\sin(4\pi x) - 3e^{-128\pi^2 t}\sin(8\pi x) + 2e^{-200\pi^2 t}\sin(10\pi x).$$

Now we consider the solution to the boundary and initial value problem

$$\begin{aligned}
\frac{\partial u}{\partial t} &= c^2 \frac{\partial^2 u}{\partial x^2}, \quad 0 < x < L, \\
\frac{\partial u}{\partial x}(0,t) &= 0, \quad u(L,t) = 0, \\
u(x,0) &= f(x),
\end{aligned} \quad (6.17)$$

using the method of separation of variables. Note that a homogeneous Neumann boundary condition is prescribed at $x = 0$.

Step 1: Let $u(x,t) = T(t)X(x)$ and plug its partial derivatives into the original PDE so that we can separate variables. The homogeneous boundary conditions require $X'(0) = 0$ and $X(L) = 0$. Differentiating $u(x,t) = T(t)X(x)$ with t and x respectively, we get

$$\frac{\partial u}{\partial t} = T'(t)X(x), \quad \frac{\partial u}{\partial x} = T(t)X'(x), \quad \frac{\partial^2 u}{\partial x^2} = T(t)X''(x).$$

The heat equation can be re-written as

$$T'(t)X(x) = c^2 T(t)X''(x) \implies \frac{T'(t)}{c^2 T(t)} = \frac{X''(x)}{X(x)} = -\lambda. \quad (6.18)$$

This is because in the last equality, the left hand side is a function of t while the right hand side is a function of x, which is possible only both of them are a constant independent of t and x. We can solve the eigenvalue problems either for $X(x)$ or $T(t)$. Since we know the boundary condition for $X(x)$, naturally we should solve

$$\frac{X''(x)}{X(x)} = -\lambda \quad \text{or} \quad X''(x) + \lambda X(x) = 0, \quad X'(0) = X(L) = 0 \quad (6.19)$$

first.

Step 2: Solve the eigenvalue problem. From the Sturm-Liouville eigenvalue theory, we know that $\lambda > 0$ since the $q(x)$ term in the S-L theorem is zero. Thus, the solution of $X(x)$ is

$$X(x) = C_1 \cos \sqrt{\lambda} x + C_2 \sin \sqrt{\lambda} x,$$

$$X'(x) = -C_1 \sqrt{\lambda} \sin \sqrt{\lambda} x + C_2 \sqrt{\lambda} \cos \sqrt{\lambda} x.$$

From the boundary condition $X'(0) = 0$, we get $C_2 = 0$. From the boundary condition $X(L) = 0$, we get

$$C_1 \cos \sqrt{\lambda} L = 0, \quad \Longrightarrow \quad \sqrt{\lambda} L = n\pi + \frac{\pi}{2}, \quad n = 0, 1, \cdots,$$

since $C_1 \neq 0$. The eigenvalues and their corresponding eigenfunctions are

$$\lambda_n = \left(\frac{(2n+1)\pi}{2L}\right)^2, \quad X_n(x) = \cos \frac{(2n+1)\pi x}{2L}, \quad n = 0, 1, 2, \cdots.$$

Now we solve for $T(t)$ using

$$T'(t) + c^2 \lambda_n T(t) = 0. \tag{6.20}$$

The solution (not an eigenvalue problem anymore since we have already known λ_n) of $T(t)$ is

$$T_n(t) = a_n e^{-c^2 \lambda_n t} = b_n e^{-c^2 \left(\frac{(2n+1)\pi}{2L}\right)^2 t}.$$

Putting $X_n(x)$ and $T_n(t)$ together, we get a normal mode solution for each n,

$$u_n(x,t) = a_n \cos \frac{(2n+1)\pi x}{2L} e^{-c^2 \left(\frac{(2n+1)\pi}{2L}\right)^2 t}, \tag{6.21}$$

which satisfy the PDE, the boundary conditions, but not the initial condition.

Step 3: Put all the normal mode solutions together to get the series solution. The coefficients are obtained from the orthogonal expansion of the initial condition.

Thus, the solution to the initial and boundary value problem of the 1D wave equation can be written as

$$u(x,t) = \sum_{n=0}^{\infty} a_n \cos \frac{(2n+1)\pi x}{2L} e^{-c^2 \left(\frac{(2n+1)\pi}{2L}\right)^2 t} \tag{6.22}$$

which satisfies the PDE and the boundary conditions. The coefficients of b_n are determined from the initial conditions $u(x,0)$,

$$u(x,0) = \sum_{n=0}^{\infty} a_n \cos \frac{(2n+1)\pi x}{2L} \quad \Longrightarrow \quad a_n = \frac{2}{L} \int_0^L f(x) \cos \frac{(2n+1)\pi x}{2L} dx,$$

which is a Fourier expansion of the initial condition $u(x,0) = f(x)$ on $(0, L)$.

6.3. Series solution to 1D heat equations with various BC's

> **Solutions to the 1D heat equation BVP with a homogeneous Neumann BC have the following uniform form,**
>
> $$u(x,t) = \sum_{n=0}^{\infty} a_n \cos \frac{(2n+1)\pi x}{2L} e^{-c^2 \left(\frac{(2n+1)\pi x}{2L}\right)^2 t},$$
>
> $$a_n = \frac{2}{L} \int_0^L f(x) \cos \frac{(2n+1)\pi x}{2L} dx.$$
>
> (6.23)

Example 6.3. *Find the series solution for the heat equation of initial and boundary value problem,*

$$\frac{\partial u}{\partial t} = \frac{\partial^2 u}{\partial x^2}, \quad 0 < x < 3,$$

$$\frac{\partial u}{\partial x}(0,t) = 0, \quad u(3,t) = 0,$$

$$u(x,0) = \begin{cases} 1 & \text{if } 0 \leq x \leq 1, \\ 1 - \dfrac{x}{2} & \text{if } 1 \leq x \leq 2, \\ 0 & \text{if } 2 < x \leq 3. \end{cases}$$

Solution: The example can be considered as the temperature distribution when a rot was heated in some parts. Note that in this example, we have $L = 3$, $c = 1$, and a homogeneous Neumann boundary condition at $x = 0$. The computation of the coefficients a_n is somewhat complicated,

$$a_n = \frac{2}{3} \int_0^3 f(x) \cos \frac{(2n+1)\pi x}{6} dx$$

$$= \frac{2}{3} \left(\int_0^1 \cos \frac{(2n+1)\pi x}{6} dx + \int_1^2 \left(1 - \frac{x}{2}\right) \cos \frac{(2n+1)\pi x}{6} dx \right)$$

$$= \frac{2\sqrt{3}\sin\frac{n\pi}{3} + 2\cos\frac{n\pi}{3}}{(2n+1)\pi} + \frac{\left(36\sqrt{3}\sin\frac{n\pi}{3} + 18\sqrt{3} - (3+6n)\pi\right)\cos\frac{n\pi}{3}}{3(2n+1)^2\pi^2}$$

$$+ \frac{18 - 36\cos\left(\frac{n\pi}{3}\right)^2 - \left(18 + (6n+3)\sqrt{3}\pi\right)\sin\frac{n\pi}{3}}{3(2n+1)^2\pi^2}.$$

The computation has been verified by Maple. Thus, the series solution is

$$u(x,t) = \sum_{n=0}^{\infty} a_n \cos \frac{(2n+1)\pi x}{6} e^{-\left(\frac{(2n+1)\pi x}{6}\right)^2 t}.$$

In Figure 6.1, we show plots of the initial condition and several partial sums of its series expansion with $N = 1$, $N = 5$, and $N = 175$. The series approximates the function very well when N is large enough but oscillates at the discontinuity $x = 1$, which is the Gibb's phenomena. However, for heat equations, the oscillations will soon be dampened and the solution becomes smooth. In the Maple file, one can use the animation feature to see the evolution of the solution. Note that due to the Neumann boundary condition at $x = 0$, the solution at $x = 0$ is not fixed and moves to the steady state solution $\lim_{t \to \infty} u(x,t) = 0$ gradually as the rest part of the solution.

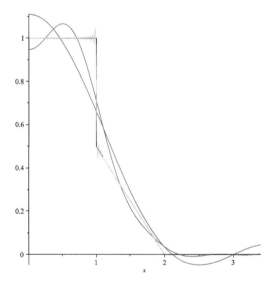

Figure 6.1. *Plots of the initial condition, three partial sums, $N = 1, 5, 175$ of the series expansion of the initial condition.*

6.3.1 Summary of series solutions of 1D heat equations with homogeneous linear BC's

Similar to one-dimensional wave equations, we can summarize the series solutions to 1D heat equations.

6.3. Series solution to 1D heat equations with various BC's

Series solutions of 1D heat equations with homogeneous linear BC's:

$$\frac{\partial u}{\partial t} = c^2 \frac{\partial^2 u}{\partial x^2}, \qquad u(x,0) = f(x), \qquad 0 < x < L,$$

with homogeneous Dirichlet, Neumann, and Robin boundary conditions below.

$$u(x,t) = \sum_{n=1 \text{ or } 0}^{\infty} a_n X_n\left(\frac{\alpha_n \pi x}{L}\right) e^{-\left(\frac{\alpha_n \pi c}{L}\right)^2 t}, \tag{6.24}$$

where

$$a_n = \frac{2}{L} \int_0^L f(x) X_n\left(\frac{\alpha_n \pi x}{L}\right) dx. \tag{6.25}$$

Dirichlet-Dirichlet: $u(0,t) = u(L,t) = 0$. We have $X_n\left(\frac{\alpha_n \pi x}{L}\right) = \sin\frac{n\pi x}{L}$.

Dirichlet-Neumann: $u(0,t) = 0$, $\frac{\partial u}{\partial x}(L,t) = 0$. We have $X_n\left(\frac{\alpha_n \pi x}{L}\right) = \sin\frac{(\frac{1}{2}+n)\pi x}{L}$.

Neumann-Dirichlet: $\frac{\partial u}{\partial x}(0,t) = 0$, $u(L,t) = 0$. We have $X_n\left(\frac{\alpha_n \pi x}{L}\right) = \cos\frac{(\frac{1}{2}+n)\pi x}{L}$.

Neumann-Neumann: Any constant is a solution and the solution is not unique.

Dirichlet-Robin: $u(0,t) = 0$, $u(L,t) + \frac{\partial u}{\partial x}(L,t) = 0$. We have $X_n\left(\frac{\alpha_n \pi x}{L}\right) = \sin\frac{\alpha_n \pi x}{L}$, where we do not have analytic expressions for α_n.

For Robin-Dirichlet, or Neumann-Robin, or other linear boundary conditions, we generally do not have an analytic form for the eigenvalues and eigenfunctions. Thus, there are no analytic series solutions available in general.

6.3.2 Steady state solutions of 1D heat equations of BVPs

A steady state solution to a differential equation is a function independent of time t that satisfies the following:

- It is a solution to the differential equation;

- It satisfies the boundary condition but it is independent of the initial condition;

- It is independent of time t, i.e., $\dfrac{\partial u}{\partial t} = 0$.

A steady state solution is the result of long time behavior of the solution. Note that, not all problems have a steady state solution.

Example 6.4. *Find the steady state solution of the following if it exists,*

$$\frac{\partial u}{\partial t} = c^2 \frac{\partial^2 u}{\partial x^2}, \qquad 0 < x < L,$$

$$u(0,t) = T_1, \qquad u(L,t) = T_2,$$

$$u(x,0) = f(x), \qquad 0 < x < L.$$

Solution: The steady state solution denoted as $u_s(x)$ is the solution to the following problem ($u(x,t) \implies u_s(x)$),

$$0 = c^2 \frac{d^2 u_s}{dx^2}, \qquad 0 < x < L,$$

$$u_s(0) = T_1, \qquad u_s(L) = T_2.$$

The general solution is $u_s(x) = C_1 + C_2 x$. The boundary condition $u_s(0) = T_1$ leads to $C_1 = T_1$ and $u_s(L) = T_2$ leads to the steady state solution

$$u_s(x) = T_1 + \frac{T_2 - T_1}{L} x.$$

One application of a steady state solution is to transform non-homogeneous boundary conditions to homogeneous ones. If we want to solve the 1D heat equation above for anytime (not just long term behavior), we can define $w(x,t) = u(x,t) - u_s(x)$. Then, $w(x,t)$ satisfies the homogeneous boundary conditions and is the solution to the following

$$\frac{\partial w}{\partial t} = c^2 \frac{\partial^2 w}{\partial x^2}, \qquad 0 < x < L,$$

$$w(0,t) = 0, \qquad w(L,t) = 0,$$

$$w(x,0) = f(x) - u_s(x), \qquad 0 < x < L.$$

Once we have solved $w(x,t)$, we get back the solution $u(x,t) = w(x,t) + u_s(x)$.

Example 6.5. *Solve the heat equation of the following*

$$\frac{\partial u}{\partial t} = 2\frac{\partial^2 u}{\partial x^2}, \quad 0 < x < 3,$$

$$u(0,t) = 10, \quad u(3,t) = 40,$$

$$u(x,0) = 25, \quad 0 < x < 3.$$

Solution: In this example, we have $u_s(x) = 10 + 10x$. The function $w(x,t)$ satisfies the heat equation with the homogeneous boundary condition and the initial condition $w(x,0) = u(x,0) - u_s(x) = 25 - 10 - 10x = 15 - 10x$. From the formula in (4.37), we have

$$w(x,t) = \sum_{n=1}^{\infty} b_n \sin\frac{n\pi x}{3} e^{-2(\frac{n\pi}{3})^2 t},$$

where the coefficients are given

$$b_n = \frac{2}{3}\int_0^3 (15 - 10x)\sin\frac{n\pi x}{3}\,dx = \frac{30}{n\pi}(\cos n\pi - 1), \quad n = 1, 2, \cdots.$$

Thus, the solution to the original problem is

$$u(x,t) = 10 + 10x + \sum_{n=1}^{\infty} \frac{30}{n\pi}(\cos n\pi - 1) e^{-2(\frac{n\pi}{3})^2 t} \sin\frac{n\pi x}{3}.$$

Note that not all the time dependent problems have steady state solutions. For example, for the heat equation with $u(0,t) = \sin t$, the boundary condition depends on t and hence there is no steady state solutions.

6.4 Two-dimensional Laplace equations of BVPs on rectangles

In this section, we consider the series solution to Laplace or Poisson equations on a rectangular domain $\mathcal{R}$,

$$\frac{\partial^2 u}{\partial x^2} + \frac{\partial^2 u}{\partial y^2} = 0, \quad \text{or} \quad u_{xx} + u_{yy} = 0, \quad (x,y) \in \mathcal{R}, \tag{6.26}$$

$$u(x,y)\Big|_{\partial \mathcal{R}} = w(x,y), \quad \text{or} \quad \frac{\partial u}{\partial n}(x,y)\Big|_{\partial \mathcal{R}} = g(x,y), \tag{6.27}$$

or other boundary conditions, where $\frac{\partial u}{\partial n}(x,y)$ is the directional derivative of $u(x,y)$ along the outer normal direction $\mathbf{n}$ ($|\mathbf{n}| = 1$). We use $\mathcal{R}$ to represent the rectangular domain, while $\partial \mathcal{R}$ as the boundary of the rectangle. The partial differential equation is second order, constant coefficients, homogeneous, linear PDE in two space dimensions. It is classified as an elliptic PDE and the solution is a global one which means that solution depends on the solution in the entire domain.

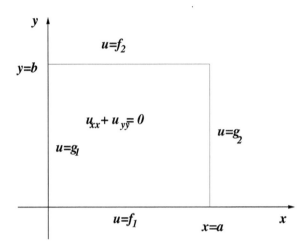

Figure 6.2. *A diagram of a Laplace equation defined on a rectangular domain with a Dirichlet boundary condition.*

We can use the gradient operator $\nabla u = [\frac{\partial}{\partial x}, \frac{\partial}{\partial y}]^T$ to represent the Laplace/Poisson equation in any (space) dimensions using $\nabla^2 u = 0$, or $\Delta u = 0$, for example, in two dimensions, we have

$$\nabla^2 u = \nabla \cdot (\nabla u) = \begin{bmatrix} \frac{\partial}{\partial x} \\ \frac{\partial}{\partial y} \end{bmatrix}^T \cdot \begin{bmatrix} \frac{\partial u}{\partial x} \\ \frac{\partial u}{\partial y} \end{bmatrix} = \Delta u = \frac{\partial^2 u}{\partial x^2} + \frac{\partial^2 u}{\partial y^2}. \qquad (6.28)$$

The operator $\nabla^2 = \Delta$ is called the Laplace operator, where $\mathbf{x}^T$ denotes the transpose of the vector $\mathbf{x}$. Note that the solution of a Laplace equation can be considered as the steady state solution of a 2D heat equation (or a wave equation)

$$\frac{\partial u}{\partial t} = \frac{\partial^2 u}{\partial x^2} + \frac{\partial^2 u}{\partial y^2}, \quad (x,y) \in \mathcal{R}, \qquad (6.29)$$

$$u(x,y,t)\Big|_{\partial \mathcal{R}} = w(x,y), \quad \text{or} \quad \frac{\partial u}{\partial n}(x,y,t)\Big|_{\partial \mathcal{R}} = g(x,y), \qquad (6.30)$$

$$u(x,y,0) = f(x,y) \qquad (6.31)$$

6.4. Two-dimensional Laplace equations of BVPs on rectangles

for an arbitrary two-dimensional function $f(x, y)$. A Neumann boundary condition means that the directional derivative is prescribed along the normal direction **n** pointing to outside of the domain following the right hand side rule. For example, for the diagram in Figure 6.2, at $x = 0$, the left boundary, a Neumann boundary condition means that $\frac{\partial u}{\partial n} = -\frac{\partial u}{\partial x}$ is given; while at $x = a$, the right boundary, a Neumann boundary condition means that $\frac{\partial u}{\partial n} = \frac{\partial u}{\partial x}$ is given. Applications of Laplace equations can be found in potential flows, ideal flows, potential of electro-magnetics. A conservative vector field satisfying $div(\mathbf{u}) = curl \times \mathbf{u} = 0$ can be represented as a potential of a scalar function, $\mathbf{u} = \nabla\varphi$ and $\Delta\varphi = 0$. An example is the Newtonian gravitational field. We can only solve one scalar equation instead of three equations of a conservative vector field.

It is easy to check that $u(x, y) = \frac{1}{2\pi} \log \sqrt{x^2 + y^2}$ is a solution to the 2D Laplace equation. It is called the fundamental solution for a 2D Laplace equation, which corresponds to a point source (charge) at $(0,0)$. In three dimension, the fundamental solution is $u(x,y,z) = \frac{1}{r} = \frac{1}{\sqrt{x^2+y^2+z^2}}$. The fundamental solution satisfies the PDE but not to boundary conditions in general.

To use the method of separation of variables, we wish to have at least two homogeneous boundary conditions. Since the problem is linear, we can split the problem into four sub-problems, see Figure 6.3 for an illustration. The final solution will be the sum of the solutions of the sub-problems.

We solve one of the problems in Figure 6.3, the top-right one,

$$\frac{\partial^2 u}{\partial x^2} + \frac{\partial^2 u}{\partial y^2} = 0, \quad (x, y) \in \mathcal{R}, \tag{6.32}$$

$$u(x, b) = f_2(x), \quad u(x, 0) = 0, \quad u(0, y) = 0, \quad u(a, y) = 0. \tag{6.33}$$

We set $u(x, y) = X(x)Y(y)$ and separate the variables to get

$$\frac{X''}{X} = -\frac{Y''}{Y} = -\lambda. \tag{6.34}$$

We inspect the two homogeneous boundary conditions that are $X(0) = 0$ and $X(a) = 0$. Thus, we solve the Sturm-Liouville eigenvalue problem for $X(x)$ to get

$$\lambda_n = \left(\frac{n\pi}{a}\right)^2, \quad X_n(x) = \sin\frac{n\pi x}{a}, \quad n = 1, 2, \cdots. \tag{6.35}$$

Next, we solve $Y(y)$ from $-\frac{Y''}{Y} = \lambda_n = \left(\frac{n\pi}{a}\right)^2$ for each n which is not an S-L eigenvalue problem anymore. The solution can be expressed as

$$Y_n(y) = b_n e^{-\frac{n\pi x}{a}} + b_n^* e^{\frac{n\pi x}{a}}, \tag{6.36}$$

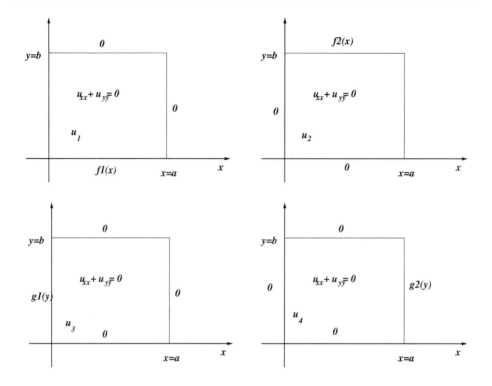

Figure 6.3. *A diagram of the decomposition of the solution to a Laplace equation on a rectangular domain into four sub-problems.*

or the hyperbolic sine and cosine functions

$$Y_n(y) = B_n \sinh \frac{n\pi y}{a} + B_n^* \cosh \frac{n\pi y}{a}. \tag{6.37}$$

The hyperbolic sine and cosine functions are defined by

$$\sinh x = \frac{e^x - e^{-x}}{2}, \qquad \cosh = \frac{e^x + e^{-x}}{2}, \tag{6.38}$$

respectively. They are linear independent since the Wronskian

$$\det \begin{pmatrix} \sinh x & \cosh x \\ \cosh x & \sinh x \end{pmatrix} = -1 \neq 0 \quad \text{for any } x. \tag{6.39}$$

The hyperbolic cosine and sine functions have similar properties as sine and cosine functions such as $\sinh(0) = 0$, $\cosh(0) = 1$, $\sinh' x = \cosh x$, $\cosh' x = -\sinh x$ etc. Thus, it is easier and similar to the discussions in previous chapters using hyperbolic sine and cosine functions. From $Y(0) = 0$, we get $B_n^* = 0$ and we can write the

6.4. Two-dimensional Laplace equations of BVPs on rectangles

solution to the Laplace equation as

$$u(x,y) = \sum_{n=1}^{\infty} B_n \sin \frac{n\pi x}{a} \sinh \frac{n\pi y}{a}.$$

Plug the non-homogenous boundary condition along $y = b$, we get

$$u(x,b) = \sum_{n=1}^{\infty} B_n \sin \frac{n\pi x}{a} \sinh \frac{n\pi b}{a} = f_2(x).$$

Note that $\sinh \frac{n\pi b}{a}$ are constants and the set $\{\sin \frac{n\pi x}{a} \sinh \frac{n\pi b}{a}\}$ is still an orthogonal set. From the orthogonal function expansion, we obtain the formula for the coefficients,

$$B_n \sinh \frac{n\pi b}{a} = \frac{2}{a} \int_0^a f_2(x) \sin \frac{n\pi x}{a} dx,$$

$$\implies B_n = \frac{2}{a \sinh \frac{n\pi b}{a}} \int_0^a f_2(x) \sin \frac{n\pi x}{a} dx.$$

Example 6.6. *Solve the Laplace equation*

$$\frac{\partial^2 u}{\partial x^2} + \frac{\partial^2 u}{\partial y^2} = 0, \quad 0 < x, y < 1, \quad (\text{a unit square in } x\text{-}y \text{ plane}),$$

$$u(x,1) = x(1-x), \quad u(x,y) = 0, \quad \text{on other three boundaries.}$$

Solution: In this example, $a = 1, b = 1$, we have

$$B_n = \frac{2}{a \sinh \frac{n\pi b}{a}} \int_0^a f_2(x) \sin \frac{n\pi x}{a} dx = \frac{2}{\sinh n\pi} \int_0^1 x(1-x) \sin n\pi x\, dx$$

$$= \frac{2}{n\pi \sinh n\pi}(-\cos n\pi x)x(1-x)\Big|_0^1 + \frac{2}{n\pi \sinh n\pi}\int_0^1 (1-2x)\cos n\pi x\, dx$$

$$= \frac{2}{(n\pi)^2 \sinh n\pi}(1-2x)\sin n\pi x \Big|_0^1 - \frac{2}{(n\pi)^2 \sinh n\pi}\int_0^1 (-2)\sin n\pi x\, dx$$

$$= \frac{4}{(n\pi)^3 \sinh n\pi}(-\cos n\pi x)\Big|_0^1 = -\frac{4}{(n\pi)^3 \sinh n\pi}(\cos n\pi - 1)$$

$$= \frac{8}{(2k-1)^3 \pi^3 \sinh(2k-1)\pi}, \quad k = 1, 2, \cdots.$$

The solution then is, see also Figure 6.4 and the Maple file Laplace.mw,

$$u(x,y) = \sum_{n=1}^{\infty} \frac{8}{(2n-1)^3 \pi^3} \frac{\sin(2n-1)\pi x \, \sinh((2n-1)\pi y)}{\sinh(2n-1)\pi}$$

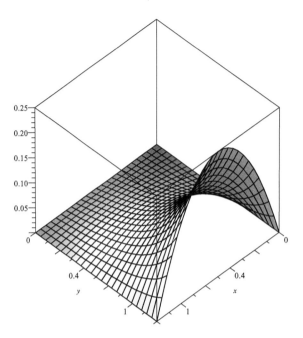

Figure 6.4. *Plot of the partial sum $S_{20}(x,y)$ of the series solution.*

How do we find the solution for the first case, the top-left diagram in Figure 6.3, i.e., $u_1(x,0) = f_1(x)$ and $u_1(x,y) = 0$ on other three boundaries? We can repeat the method of separation of variables; or we can change the problem to one that we have already solved.

Let $\bar{y} = b - y$, $\bar{x} = x$, then

$$u_1(x,y) = u_1(\bar{x}, b - \bar{y}) \stackrel{define}{=} \bar{u}_1(\bar{x}, \bar{y}).$$

Then, we have the following

$$\frac{\partial^2 \bar{u}_1}{\partial \bar{x}^2} = \frac{\partial^2 u_1}{\partial x^2}, \quad \frac{\partial \bar{u}_1}{\partial \bar{y}} = -\frac{\partial u_1}{\partial y}, \quad \frac{\partial^2 \bar{u}_1}{\partial \bar{y}^2} = \frac{\partial^2 u_1}{\partial y^2}$$

$$\frac{\partial^2 \bar{u}_1}{\partial \bar{x}^2} + \frac{\partial^2 \bar{u}_1}{\partial \bar{y}^2} = \frac{\partial^2 u_1}{\partial x^2} + \frac{\partial^2 u_1}{\partial y^2} = 0,$$

$$\bar{u}_1(0, \bar{y}) = u_1(0, b - y) = 0, \quad \bar{u}_1(a, \bar{y}) = u_1(a, b - y) = 0, \quad \bar{u}_1(\bar{x}, 0) = u_1(x, b) = 0,$$

$$\bar{u}_1(\bar{x}, b) = u_1(x, 0) = f_1(x).$$

We apply the previous solution formula to get

$$\bar{u}_1(\bar{x}, \bar{y}) = \sum_{n=0}^{\infty} A_n \sin \frac{n\pi \bar{x}}{a} \sinh \frac{n\pi \bar{y}}{a}, \quad A_n = \frac{2}{a \sinh \frac{n\pi b}{a}} \int_0^a f_1(x) \sin \frac{n\pi x}{a} dx.$$

We switch to the original coordinates to get

$$u(x,y) = \sum_{n=1}^{\infty} A_n \sin\frac{n\pi x}{a} \sinh\frac{n\pi(b-y)}{a}.$$

For the non-homogeneous part along the boundaries $x = 0$ and $x = a$, we can use the symmetry arguments by switching x with y, and a with b, we can get a formula for the entire problem:

Solution to Laplace equation on a rectangle

$$\begin{aligned} u(x,y) &= \sum_{n=1}^{\infty} A_n \sin\frac{n\pi x}{a} \sinh\frac{n\pi(b-y)}{a} + \sum_{n=1}^{\infty} B_n \sin\frac{n\pi x}{a} \sinh\frac{n\pi y}{a} \\ &+ \sum_{n=1}^{\infty} C_n \sin\frac{n\pi y}{b} \sinh\frac{n\pi(a-x)}{b} + \sum_{n=1}^{\infty} D_n \sin\frac{n\pi y}{b} \sinh\frac{n\pi x}{b}, \end{aligned} \tag{6.40}$$

where the coefficients are

$$A_n = \frac{2}{a \sinh\frac{n\pi b}{a}} \int_0^a f_1(x) \sin\frac{n\pi x}{a} dx, \quad B_n = \frac{2}{a \sinh\frac{n\pi b}{a}} \int_0^a f_2(x) \sin\frac{n\pi x}{a} dx,$$

$$C_n = \frac{2}{b \sinh\frac{n\pi a}{b}} \int_0^b g_1(y) \sin\frac{n\pi y}{b} dx, \quad D_n = \frac{2}{b \sinh\frac{n\pi a}{b}} \int_0^b g_2(x) \sin\frac{n\pi y}{b} dx.$$

6.5 Double series solutions for 2D wave equations of BVPs*

For two- and three-dimensional problems, the method of separation of variables leads to double or triple series solutions, respectively. We use a two-dimensional wave equation example to illustrate the process.

Consider a wave equation on a rectangular domain with homogeneous boundary condition:

$$\frac{\partial^2 u}{\partial t^2} = c^2 \left(\frac{\partial^2 u}{\partial x^2} + \frac{\partial^2 u}{\partial y^2} \right), \quad (x,y) \in \mathcal{R}, \quad t > 0$$

$$u(x,y,t)\Big|_{\partial \mathcal{R}} = 0, \quad \mathcal{R} = (0, a) \times (0, b) \tag{6.41}$$

$$u(x,y,0) = f(x,y), \quad \frac{\partial u}{\partial t}(x,y,0) = g(x,y),$$

where we use $\mathcal{R}$ to represent the rectangular domain, and $\partial \mathcal{R}$ to represent its boundary. To solve the problem using the method of separation of variables, we perform the usual steps.

Step 1: Let $u(x,y,t) = T(t)X(x)Y(y)$ and plug its partial derivatives into the original PDE so that we can separate variables. The homogeneous boundary conditions require $X(0) = X(a) = 0$ and $Y(0) = Y(b) = 0$. Differentiating $u(x,y,t) = T(t)X(x)Y(y)$ with respect to t, x, and y, we get

$$\frac{\partial u}{\partial t} = T'(t)X(x)Y(y), \quad \frac{\partial^2 u}{\partial t^2} = T''(t)X(x)Y(y);$$

$$\frac{\partial u}{\partial x} = T(t)X'(x)Y(y), \quad \frac{\partial^2 u}{\partial x^2} = T(t)X''(x)Y(y);$$

$$\frac{\partial u}{\partial y} = T(t)X(x)Y'(y), \quad \frac{\partial^2 u}{\partial y^2} = T(t)X(x)Y''(y).$$

The wave equation can be re-written as

$$T''(t)X(x)Y(y) = c^2 T(t)\left(X''(x)Y(y) + Y''(y)X(x)\right),$$
$$\implies \frac{T''(t)}{c^2 T(t)} = \frac{X''(x)}{X(x)} + \frac{Y''(y)}{Y(y)} = -\nu. \tag{6.42}$$

This is because in the last equality, the left hand side is a function of t while the right hand side is a function of x and y, which is possible only both of them are a constant independent of t and x and y.

We can separate variables further since $\frac{X''(x)}{X(x)}$ is a function of x, and $\frac{Y''(x)}{Y(y)}$ is a function of y to write

$$\frac{X''(x)}{X(x)} = -\frac{Y''(x)}{Y(y)} - \nu = -\mu. \tag{6.43}$$

We get three ordinary differential equations for $X(x)$, $Y(y)$, and $T(t)$. Since we know the boundary condition for $X(x)$ and $Y(y)$, naturally we should solve them first

$$\frac{X''(x)}{X(x)} = -\mu \quad \text{or} \quad X''(x) + \mu X(x) = 0, \quad X(0) = 0, \quad X(a) = 0. \tag{6.44}$$

Step 2: Solve the eigenvalue problems for $X(x)$ and $Y(y)$. From the Sturm-Liouville eigenvalue theory, we know that the solution is

$$X''(x) = C_1 \cos \sqrt{\mu}\, x + C_2 \sin \sqrt{\mu}\, x.$$

6.5. Double series solutions for 2D wave equations of BVPs*

From the boundary condition $X(0) = 0$, we get $C_1 = 0$. From the boundary condition $X(a) = 0$, we get

$$C_2 \sin \sqrt{\mu} a = 0, \quad \Longrightarrow \quad \sqrt{\mu} a = m\pi, \quad m = 1, 2, \cdots,$$

since $C_2 \neq 0$. The eigenvalues and their corresponding eigenfunctions are

$$\mu_m = \left(\frac{m\pi}{a}\right)^2, \quad X_m(x) = \sin \frac{m\pi x}{a}, \quad m = 1, 2, \cdots,.$$

Similarly, we set $\frac{Y''}{Y} = -\nu + \mu = -\gamma$ along with $Y(0) = Y(b) = 0$, to obtain

$$\gamma_n = \left(\frac{n\pi y}{b}\right)^2, \quad Y_n(y) = \sin \frac{n\pi y}{b}, \quad n = 1, 2, \cdots.$$

Finally, we solve for $T(t)$ using

$$\frac{T''(t)}{c^2 T(t)} = -\nu = -\mu - \gamma = -\left(\frac{m\pi}{a}\right)^2 - \left(\frac{n\pi}{a}\right)^2, \quad m, n = 1, 2, \cdots,.$$

The solution $T(t)$ is (not an eigenvalue problem anymore since we have already known $\nu_{mn} = \left(\frac{m\pi}{a}\right)^2 + \left(\frac{n\pi}{a}\right)^2$).

$$T_{mn}(t) = B_{mn} \cos(\nu_{mn} ct) + B^*_{mn} \sin(\nu_{mn} ct).$$

Put $X_m(x)$, $Y_n(y)$, and $T_{mn}(t)$ together, we get a normal mode solution

$$u_{mn}(x, y, t) = \sin \frac{m\pi x}{a} \sin \frac{n\pi y}{b} \left(B_{mn} \cos(\nu_{mn} ct) + B^*_{mn} \sin(\nu_{mn} ct) \right) \quad (6.45)$$

which satisfy the PDE, the boundary conditions, but not the initial conditions.

Step 3: Put all the normal mode solutions together to get the series solution. The coefficients are obtained from the orthogonal expansion of the initial conditions. The solution to the initial and boundary value problem of the 2D wave equation can be written as

$$u(x, y, t) = \sum_{m=1}^{\infty} \sum_{m=1}^{\infty} \sin \frac{m\pi x}{a} \sin \frac{n\pi y}{b} \left(B_{mn} \cos(\nu_{mn} ct) + B^*_{mn} \sin(\nu_{mn} ct) \right),$$

which satisfies the PDE and the boundary conditions. The coefficients of B_{mn} and B^*_{mn} are determined from the initial conditions $u(x, y, 0)$ and $u_t(x, y, 0)$. If we plug $t = 0$ into the series solution, then we get

$$u(x, y, 0) = \sum_{m=1}^{\infty} \sum_{m=1}^{\infty} \sin \frac{m\pi x}{a} \sin \frac{n\pi y}{b} B_{mn} = f(x, y).$$

From this we obtain

$$B_{mn} = \frac{\int_0^a \int_0^b f(x,y) \sin\frac{m\pi x}{a} \sin\frac{n\pi y}{b} dxdy}{\int_0^a \int_0^b \sin^2\frac{m\pi x}{a} \sin^2\frac{n\pi y}{b} dxdy}$$

$$= \frac{4}{ab} \int_0^a \int_0^b f(x,y) \sin\frac{m\pi x}{a} \sin\frac{n\pi y}{b} dxdy,$$

since the denominator can be obtained analytically.

Similarly, we get the formula for B_{mn}^*

$$B_{mn}^* = \frac{4}{abc\,\nu_{mn}} \int_0^a \int_0^b g(x,y) \sin\frac{m\pi x}{a} \sin\frac{n\pi y}{b} dxdy,$$

where again $\nu_{mn} = \sqrt{\left(\frac{m\pi}{a}\right)^2 + \left(\frac{n\pi}{a}\right)^2}$.

In a similar way, we can use the method of separation of variables to solve three-dimensional Poisson equations or wave equations, which will lead to triple series solutions. Alternatively, we can solve those equations using some numerical methods, see for example, Chapter 8 which may be simpler.

6.6 Method of separation of variables for PDEs of BVPs in polar coordinates

In many applications it is preferable to use polar coordinates (two space dimensions) or cylindrical/spherical coordinates (three space dimensions), especially when we deal with circles, annuli, *etc.*, see Figure 6.5 for an illustration. Often we can solve a two- or three-dimensional problem using one-dimensional settings if the problem possesses axial-symmetry. How will partial differential equations be changed using the polar/cylindrical coordinates? We know that in polar coordinates

$$x = r\cos\theta, \quad y = r\sin\theta, \quad r = \sqrt{x^2 + y^2}, \quad \theta = \arctan(y/x), \quad (6.46)$$

where θ is the angle between the x-axis and the ray $\overrightarrow{OX}$, where $\mathbf{X} = (x,y)$ and O is the origin in the two-dimensional x-y plane. For a function $u(x,y)$, we can represent the function and its partial derivatives using (r,θ),

$$u(x,y) = u(r\cos\theta, r\sin\theta) = \bar{u}(r,\theta). \quad (6.47)$$

6.6. Method of separation of variables for PDEs of BVPs in polar coordinates

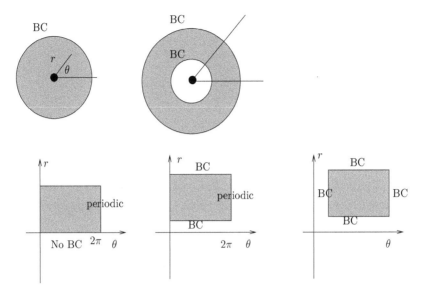

Figure 6.5. *Diagrams of domains and boundary conditions that may be better solved in polar coordinates. Top diagrams are domains in x-y plane, while the bottom diagrams are domains in r-θ plane.*

For simplicity, we often omit the bar if there is no confusion occurring. Next, we replace the partial derivatives in terms of (r, θ) as well using the chain rule to obtain

$$\frac{\partial u}{\partial x} = \frac{\partial u}{\partial r}\frac{\partial r}{\partial x} + \frac{\partial u}{\partial \theta}\frac{\partial \theta}{\partial x},$$

$$\frac{\partial u}{\partial y} = \frac{\partial u}{\partial r}\frac{\partial r}{\partial y} + \frac{\partial u}{\partial \theta}\frac{\partial \theta}{\partial y}.$$

From $r = \sqrt{x^2 + y^2}$, $\theta = \arctan(y/x)$, we also have

$$\frac{\partial r}{\partial x} = \frac{2x}{2\sqrt{x^2 + y^2}} = \frac{r\cos\theta}{r} = \frac{x}{r},$$

$$\frac{\partial \theta}{\partial x} = \frac{1}{1 + (\frac{y}{x})^2}\left(-\frac{y}{x^2}\right) = \frac{-y}{x^2 + y^2} = -\frac{y}{r^2} = -\frac{r\sin\theta}{r^2}.$$

Thus, we get

$$\frac{\partial u}{\partial x} = \frac{\partial u}{\partial r}\cos\theta + \frac{\partial u}{\partial \theta}\left(-\frac{r\sin\theta}{r^2}\right) = \frac{\partial u}{\partial r}\frac{x}{r} - \frac{\partial u}{\partial \theta}\frac{y}{r^2},$$

$$\frac{\partial u}{\partial y} = \frac{\partial u}{\partial r}\frac{y}{r} + \frac{\partial u}{\partial \theta}\frac{x}{r^2} = \frac{\partial u}{\partial r}\sin\theta + \frac{\partial u}{\partial \theta}\frac{\cos\theta}{r},$$

$$\frac{\partial^2 u}{\partial x^2} = \frac{\partial}{\partial x}\left(\frac{\partial u}{\partial r}\frac{x}{r} - \frac{\partial u}{\partial \theta}\frac{y}{r^2}\right)\cdots$$

The derivations are long and tedious. Fortunately, for most practical partial differential equations and/or vector relations, we can find the conversions through mathematical handbooks or online tools. The Laplace equation in polar coordinates in two dimensions is

$$\Delta u = \nabla^2 u = \frac{\partial^2 u}{\partial r^2} + \frac{1}{r}\frac{\partial u}{\partial r} + \frac{1}{r^2}\frac{\partial^2 u}{\partial \theta^2} = 0. \quad (6.48)$$

Note that there is no $\frac{\partial u}{\partial \theta}$ term in the expression above. We can use the dimension analysis to figure out the power coefficients of r in each of the above terms knowing that θ is a dimensionless quantity. For the radial symmetric case, that is, the solution is independent of θ, we have the simplified Laplace equation

$$\Delta u = \nabla^2 u = \frac{\partial^2 u}{\partial r^2} + \frac{1}{r}\frac{\partial u}{\partial r} = 0. \quad (6.49)$$

Series solution to the Laplace equation of BVPs on a disc

Consider the Laplace equation defined on a circle with a radius a,

$$\Delta u = \nabla^2 u = \frac{\partial^2 u}{\partial r^2} + \frac{1}{r}\frac{\partial u}{\partial r} + \frac{1}{r^2}\frac{\partial^2 u}{\partial \theta^2} = 0, \quad 0 < r < a,\ 0 < \theta < 2\pi, \quad (6.50)$$

with a Dirichlet boundary condition at the circle boundary, that is, $u(a, \theta) = f(\theta)$. Note that $r = 0$ is an interior point, not a boundary. There is no boundary condition at $r = 0$ except that the solution should be bounded. This is called the pole condition. With the method of separation of variables, we set $u(r, \theta) = R(r)\Theta(\theta)$. Following the procedure of the method of separation of variables, we obtain the following,

$$R''\Theta + \frac{1}{r}R'\Theta + \frac{1}{r^2}R\Theta'' = 0, \quad (6.51)$$

or

$$\frac{R'' + \frac{1}{r}R'}{R} + \frac{1}{r^2}\frac{\Theta''}{\Theta} = 0, \quad (6.52)$$

separate variable $\implies$ $-r^2\left(\frac{R'' + \frac{1}{r}R'}{R}\right) = \frac{\Theta''}{\Theta} = -\lambda. \quad (6.53)$

6.6. Method of separation of variables for PDEs of BVPs in polar coordinates

We have two related Sturm-Liouville problems

$$\Theta'' + \lambda\Theta = 0, \qquad \Theta(0) = \Theta(2\pi), \tag{6.54}$$

$$r^2 R'' + rR' - \lambda R = 0, \qquad 0 < r < a. \tag{6.55}$$

The boundary condition $\Theta(0) = \Theta(2\pi)$ is a periodic one. We do not know the solution of $R(0)$ and $R(a)$ except that they are bounded. Thus, we should solve the first Sturm-Liouville eigenvalue problem first. If $\lambda < 0$, we would have

$$\Theta(\theta) = C_1 e^{-\sqrt{\lambda}\theta} + C_2 e^{\sqrt{\lambda}\theta},$$

which cannot be periodic, neither the case $\lambda = 0$ for which we have $\Theta(\theta) = C_1 + C_2\theta$. Thus, we must have $\lambda > 0$ for which the solution is

$$\Theta(\theta) = C_1 \cos\sqrt{\lambda}\theta + C_2 \sin\sqrt{\lambda}\theta.$$

Apply the periodic boundary condition, we should have

$$\Theta(\theta) = C_1 \cos\sqrt{\lambda}(\theta + 2\pi) + C_2 \sin\sqrt{\lambda}(\theta + 2\pi) = C_1 \cos\sqrt{\lambda}\theta + C_2 \sin\sqrt{\lambda}\theta,$$

which leads to $2\pi\sqrt{\lambda} = 2\pi n$, $n = 0, 1, 2, \cdots$, or $\lambda_n^2 = n^2$. Note that in this case, $n = 0$ is a valid solution. The eigenfunctions then are

$$\Theta_n(\theta) = a_n \cos n\theta + b_n \sin n\theta, \qquad n = 0, 1, 2, \cdots. \tag{6.56}$$

Next, we use the second ordinary differential equation (6.55) to solve for $R(r)$ which is not an S-L eigenvalue problem since $\lambda_n = n^2$ is known for $n = 0, 1, \cdots$.

$$r^2 R'' + rR' - n^2 R = 0, \qquad 0 < r < a. \tag{6.57}$$

It is an Euler's equation, see Appendix A.4. The indicial equation is

$$\alpha(\alpha - 1) + \alpha - n^2 = 0, \tag{6.58}$$

whose solutions are

$$R(r) = C_n \left(\frac{r}{a}\right)^n + \bar{C}_n \left(\frac{r}{a}\right)^{-n}, \qquad n = 0, 1, 2, \cdots, \tag{6.59}$$

using a convenient form. Since the solution is bounded at $r = 0$, we have to have $\bar{C}_n = 0$ for $n \geq 1$. Thus, the series solution is

$$u(r, \theta) = a_0 + \sum_{n=1}^{\infty} \left(\frac{r}{a}\right)^n \left(a_n \cos n\theta + b_n \sin n\theta\right). \tag{6.60}$$

We use $\left(\frac{r}{a}\right)^n$ form instead of r^n just for convenience as we can see soon. We apply the boundary condition $(r=a)$ to get

$$u(a,\theta) = a_0 + \sum_{n=1}^{\infty}(a_n \cos n\theta + b_n \sin n\theta) = f(\theta), \qquad (6.61)$$

which is the Fourier series expansion of $f(\theta)$. The coefficients are

$$a_0 = \frac{1}{2\pi}\int_{-\pi}^{\pi} f(\theta)d\theta, \qquad a_n = \frac{1}{\pi}\int_{-\pi}^{\pi} f(\theta)\cos n\theta\, d\theta,$$

$$b_n = \frac{1}{\pi}\int_{-\pi}^{\pi} f(\theta)\sin n\theta\, d\theta, \qquad n=1,2,\cdots.$$

Thus, we have found the series solution to the Laplace equation on a disc.

Example 6.7. *Find the steady state solution of the following*

$$\frac{\partial u}{\partial t} = \Delta u, \qquad x^2 + y^2 < 1,$$

$$u(1,\theta,t) = 100 - e^{-t}, \quad BC \qquad u(r,\theta,0) = r\sin\theta, \quad IC.$$

Solution: The steady state solution is the solution to the following boundary value problem,

$$\Delta u_s = 0, \qquad x^2 + y^2 < 1,$$

$$u_s(1,\theta) = 100.$$

We can compute the coefficients of the Fourier series of $f(\theta) = 100$ to get $a_0 = 100$, $a_n = 0$ and $b_n = 0$. The Fourier series itself corresponds to $\cos(0\cdot x)$ and the steady state solution is $u_s(r,\theta) = 100$.

How about $u(1,\theta,t) = \sin 5\theta + \cos 7\theta$? The steady state solution is the normal modes solution $u_s(r,\theta) = r^5 \sin 5\theta + r^7 \cos 7\theta$.

How about $u(1,\theta,t) = 100$ if $0 < \theta < \pi$ and $u(1,\theta,t) = 0$ if $\pi < \theta < 2\pi$? We have

$$a_0 = \frac{1}{2\pi}\int_0^{\pi} 100\, d\theta = 50, \qquad a_n = \frac{1}{\pi}\int_0^{\pi} 100\cos n\theta\, d\theta = 0,$$

$$b_n = \frac{1}{\pi}\int_0^{\pi} 100\sin n\theta\, d\theta = -\frac{100}{\pi}\frac{\cos n\theta}{n}\Big|_0^{\pi} = \frac{100}{n\pi}\left(1-(-1)^n\right).$$

6.6. Method of separation of variables for PDEs of BVPs in polar coordinates

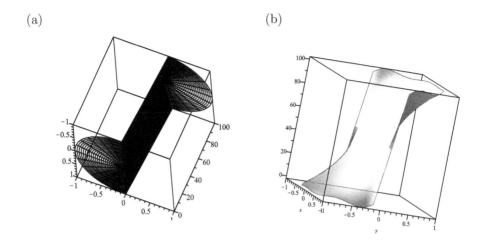

Figure 6.6. *(a): Plot of the boundary condition on the unit circle. (b): mesh plot of the partial sum (N = 40) of the series solution to the Laplace equation on the circle. Gibb's oscillations are visible due to the discontinuity at $\theta = \pi$ but the solution is smooth in the interior.*

The solution is

$$u_s(r,\theta) = 50 + \frac{100}{\pi} \sum_{n=1}^{\infty} \frac{1}{n} \left(1 - (-1)^n\right) r^n \sin n\theta$$

$$= 50 + \frac{200}{\pi} \sum_{k=1}^{\infty} \frac{r^{2k+1}}{2k+1} \sin(2k+1)\theta.$$

The series is uniformly convergent if $r \leq \alpha < 1$, but not in $(0,1)$. In Figure 6.6 (a), we show a mesh plot of the boundary condition along $r = 1$ using the polar coordinates. In Figure 6.6 (b), we show a mesh plot of the partial sum of the series solution with $N = 40$ to the Laplace equation on the circle. Gibb's oscillations are visible due to the discontinuity at $\theta = \pi$ but the solution is smooth in the interior. Solutions to Laplace equations are also called harmonic functions that are indefinitely differentiable in any interior domains.

6.7 Series solution of 2D wave equations of BVPs with a radial symmetry

A wave equation in two space dimensions in polar coordinates can be written as

$$\frac{\partial^2 u}{\partial t^2} = c^2 \left(\frac{\partial^2 u}{\partial r^2} + \frac{1}{r}\frac{\partial u}{\partial r} + \frac{1}{r^2}\frac{\partial^2 u}{\partial \theta^2} \right).$$

If the problem has the radial symmetry, then $\frac{\partial^2 u}{\partial \theta^2} = 0$. Let us consider a Dirichlet boundary condition for a wave equation on a disk with a radial symmetry,

$$\frac{\partial^2 u(r,t)}{\partial t^2} = c^2 \left(\frac{\partial^2 u}{\partial r^2} + \frac{1}{r}\frac{\partial u}{\partial r} \right), \quad 0 < r < a,$$

$$u(a,t) = 0, \qquad \text{BC at the disk, no BC at } r = 0, \tag{6.62}$$

$$u(r,0) = f(r), \quad \frac{\partial u(r,0)}{\partial t} = g(r), \qquad \text{initial conditions.}$$

We can solve the problem using the method of separation of variables as follows. First we set $u(r,t) = R(r)T(t)$; then differentiate $u(r,t)$ with respect to r and t, respectively; and plug them into the PDE to get

$$T''(t)\,R = c^2 \left(R''\,T + \frac{1}{r} R'\,T \right),$$

and separate the variables to get

$$\frac{T''}{c^2 T} = \frac{R'' + \frac{1}{r}R'}{R} = -\lambda.$$

As before, the eigenvalues of the Sturm-Liouville problems have to be positive otherwise $T(t)$ will be a constant, or go to zero or infinity. For positive eigenvalues, we know that

$$T_n(t) = A_n \cos(c\sqrt{\lambda_n}\,t) + B_n \sin(c\sqrt{\lambda_n}\,t).$$

We cannot go further from this expression. Now let us check the equation for $R(r)$

$$R'' + \frac{1}{r}R'\lambda R = 0 \quad \Longrightarrow \quad rR'' + R' + \lambda r R = 0, \quad 0 < r < a,$$

$$\Longrightarrow \quad (rR')' + (0 + \lambda r)\,R = 0, \quad R(a) = 0, \quad R(0) \text{ is unknown but bounded.}$$

This is a weighted and singular Sturm-Liouville eigenvalue problem with weight function $r \geq 0$ and a singularity at $r = 0$, which is called the pole singularity.

6.7. Series solution of 2D wave equations of BVPs with a radial symmetry

The equation can also be written as $r^2 R'' + rR' + \lambda r^2 R = 0$. To get a standard known Sturm-Liouville problem, we change the variable using $\mu = \sqrt{\lambda}\, r$ and $R(r) = R(\mu/\sqrt{\lambda}) = \bar{R}(\mu)$. The differential equation in terms of the new variable μ is

$$\mu^2 \bar{R}'' + \mu \bar{R}' + \mu^2 \bar{R} = 0, \tag{6.63}$$

which is one of the Bessel equations

$$x^2 y'' + xy' + (x^2 - p^2) y = 0, \quad \text{the } p\text{-th order Bessel equation.} \tag{6.64}$$

For the radial symmetric wave equation (6.62), we have $p = 0$, called the zeroth Bessel equation. The general solutions to the Bessel equation of order p is

$$y(x) = C_1 J_p(x) + C_2 Y_p(x), \tag{6.65}$$

where $J_p(x)$ is called the Bessel function of the first kind which is continuous in any finite interval $[0, a]$. $Y_p(x)$ is called the p-th order Bessel function of the second kind which is unbounded as $x \to 0$ corresponding to the pole singularity. In Figure 6.7, we show several plots of different Bessel functions. Figure 6.7 (a) shows three first kind of Bessel functions in $[0, 10]$. We can see that the Bessel functions are continuous everywhere including $x = 0$. The intersections of the functions and the x-axis are the eigenvalues α_n. Figure 6.7 (b) shows three second kind of Bessel functions in the interval $(0, 2]$. We can see that the Bessel functions are unbounded as x approaches the origin. The Maple commands are the following.

```
plot({BesselJ(0,x), BesselJ(1,x),BesselJ(2,x)},x=0..10,color=[red,blue,black]);
plot({BesselY(0,x), BesselY(1,x),BesselY(2,x)},x=0..2,y=-20..2,color=[red,blue,black]);
```

The general solutions of Bessel's equation (6.63) of order $p = 0$ can thus be written using (6.65) as

$$\bar{R}(\mu) = C_1 J_0(\mu) + C_2 Y_0(\mu), \quad \text{or} \quad R(r) = C_1 J_0(\sqrt{\lambda}\, r) + C_2 Y_0(\sqrt{\lambda}\, r). \tag{6.66}$$

From the differential equation theory and our knowledge on wave propagations, the solution should be bounded at the center of the disk ($r = 0$). Thus, we conclude that the coefficient $C_2 = 0$. The solution of $R(r)$ should also satisfy the boundary condition $R(a) = 0$, or $J_0(\sqrt{\lambda}\, a) = 0$. Denote the infinitely many positive zeros α_n of $J_0(\mu)$ as

$$0 < \alpha_0 < \alpha_1 < \alpha_2 \cdots < \alpha_n < \ldots < \infty, \tag{6.67}$$

which leads to the eigenvalues $\sqrt{\lambda_n} = \dfrac{\alpha_n}{a}$. The λ_n's are the eigenvalues of the original eigenfunction $R_n(r)$. Therefore, we have the solution for $R_n(r)$,

$$R_n(r) = J_0\left(\frac{\alpha_n}{a} r\right), \tag{6.68}$$

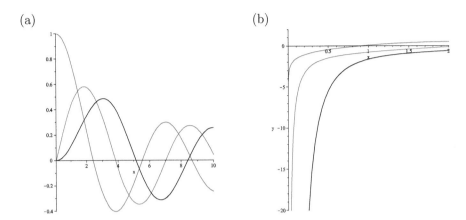

Figure 6.7. Graphs of some Bessel functions. (a): Plot of $J_0(x), J_1(x), J_2(x)$, the first three first kind of Bessel functions. (b): Plot of $Y_0(x), Y_1(x), Y_2(x)$, the first three second kind of Bessel functions.

and we know that the corresponding $T(t)$ is

$$T_n(t) = A_n \cos(c\sqrt{\lambda_n}\, t) + B_n \sin(c\sqrt{\lambda_n}\, t). \tag{6.69}$$

Finally we get the series solution to the original problem (6.62) as

$$u(r,t) = \sum_{n=1}^{\infty} \left(A_n \cos\frac{c\alpha_n t}{a} + B_n \sin\frac{c\alpha_n t}{a} \right) J_0\left(\frac{\alpha_n}{a} r\right). \tag{6.70}$$

The coefficients are determined from the initial conditions. Since $u(r,0) = f(r)$, we have

$$\sum_{n=1}^{\infty} A_n J_0\left(\frac{\alpha_n}{a} r\right) = f(r), \implies A_n = \frac{\int_0^a f(r) J_0\left(\frac{\alpha_n}{a} r\right) r\, dr}{\int_0^a J_0^2\left(\frac{\alpha_n}{a} r\right) r\, dr}. \tag{6.71}$$

Note that the weighted function r in the above integrals corresponds to the Jacobian in double integrals from Cartesian coordinates to the polar ones. The coefficients B_n are determined by taking the partial derivative of $u(r,t)$ with respect to t at $t=0$,

$$\left.\frac{\partial u}{\partial t}\right|_{t=0} = \sum_{n=1}^{\infty} \left(A_n \cdot 0 + B_n \cdot \frac{c\alpha_n}{a} \right) J_0\left(\frac{\alpha_n}{a} r\right) = g(r),$$

which leads to

$$B_n = \frac{a \int_0^a g(r) J_0\left(\frac{\alpha_n}{a} r\right) r \, dr}{c \alpha_n \int_0^a J_0^2\left(\frac{\alpha_n}{a} r\right) r \, dr}. \qquad (6.72)$$

Example 6.8. *Motion of a circular membrane with a constant initial velocity and a clamped edge.*

Consider a clamped circular elastic membrane that is initially flat. At an instance ($t = 0$), an external force such as a wind or something else triggered a uniform initial velocity, say, $\frac{\partial u}{\partial t}(r, 0)\big|_{t=0} = -100 m/sec$, find the motion (or deformation) of the membrane at any time t.

Solution: In this case, we have $u(r, 0) = 0$ and thus, $A_n = 0$. The solution is

$$u(r,t) = \sum_{n=1}^{\infty} \left(\frac{a \int_0^a g(r) J_0\left(\frac{\alpha_n}{a} r\right) r \, dr}{c \alpha_n \int_0^a J_0^2\left(\frac{\alpha_n}{a} r\right) r \, dr} \right) \sin\frac{c\alpha_n t}{a} J_0\left(\frac{\alpha_n}{a} r\right).$$

A Maple program that solves and simulates the motion with adjustable parameters of c and a is attached in this book, see Fig. 6.8. We can see animation of the motion using the following command.

```
with(plots):
animate3d([r*cos(theta),r*sin(theta),EigenfunctionExpansion],
  r=0..1,theta=0..2*Pi,t=0..40,frames=40);
```

6.8 Series solution to 3D Laplace equations of BVPs with a radial symmetry

When a domain is part of or entire a sphere, it is more convenient to use the spherical coordinates,

$$\begin{aligned} x &= r\cos\phi\sin\theta, & 0 \le \phi < 2\pi, \\ y &= r\sin\phi\sin\theta, & 0 \le \theta < \pi, \\ z &= r\cos\theta, & r = \sqrt{x^2+y^2+z^2}, \quad 0 \le r. \end{aligned} \qquad (6.73)$$

A Laplace equation in the spherical coordinates has the following form,

$$\frac{\partial^2 u}{\partial r^2} + \frac{2}{r}\frac{\partial u}{\partial r} + \frac{1}{r^2}\left(\frac{\partial^2 u}{\partial \theta^2} + \cot\theta \frac{\partial u}{\partial \theta} + \csc^2\theta \frac{\partial^2 u}{\partial \phi^2}\right) = 0. \qquad (6.74)$$

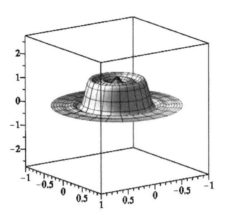

Figure 6.8. *Motion of a circular membrane with a constant initial velocity and a clamped edge.*

If the problem is radially symmetric, that is, $\frac{\partial^2 u}{\partial \phi^2} = 0$, then the Laplace equation becomes a one-dimensional problem in r that can be solved rather easily. Here we focus on the symmetry in the latitude, that is $\frac{\partial u}{\partial \phi} = 0$. We consider the Laplace equation in a sphere with a Dirichlet boundary condition,

$$\frac{\partial^2 u}{\partial r^2} + \frac{2}{r}\frac{\partial u}{\partial r} + \frac{1}{r^2}\left(\frac{\partial^2 u}{\partial \theta^2} + \cot\theta \frac{\partial u}{\partial \theta}\right) = 0, \qquad (6.75)$$
$$u(a,\theta) = f(\theta), \qquad 0 < r < a, \quad 0 \leq \theta < \pi.$$

We use the method of separation of variables to find a series solution to the boundary value problem by setting $u(r,\theta) = R(r)\Theta(\theta)$. Then, we differentiate $u(r,\theta)$ with respect to r and θ, respectively, and plug the partial derivatives into the partial differential equation to get

$$\left(R'' + \frac{2}{r}R'\right)\Theta + \frac{1}{r^2}\left(\Theta'' + \cot\theta\, \Theta'\right)R = 0,$$

and separate the variables to get

$$r^2\frac{R'' + \frac{2}{r}R'}{R} = -\frac{\Theta'' + \cot\theta\, \Theta'}{\Theta} = \lambda.$$

The equation for $R(r)$ is an Euler's equation

$$r^2 R'' + 2rR' - \lambda R = 0,$$

6.8. Series solution to 3D Laplace equations of BVPs with a radial symmetry

whose solution should be bounded at $r = 0$.[5] The indicial equation of the above Euler's equation is

$$s(s-1) + 2s - \lambda = 0, \quad \Longrightarrow \quad s_{1,2} = \frac{-1 \pm \sqrt{1+4\lambda}}{2}. \tag{6.76}$$

The solution to $R(r)$ is

$$R(r) = C_1 r^{s_1} + C_2 r^{s_2}. \tag{6.77}$$

If $1 + 4\lambda \leq 0$, then both roots s_1 and s_2 are either negative or the real part is negative that leads to an unbounded $R(r)$ at $r = 0$. Thus, the eigenvalues of λ has to be positive. That is all that we can do about $R(r)$ for now.

Next, we try to check the $\Theta(\theta)$ equation,

$$\Theta'' + \cot\theta\, \Theta' - \lambda\Theta = 0, \qquad \lambda > 0. \tag{6.78}$$

The term $\cot\theta$ is unbounded at $\theta = 0$ and $\theta = \pi$, and it is not a term that we are familiar with and can be eliminated if we set $s = \cos\theta$. Then, we have $\frac{ds}{d\theta} = -\sin\theta$, and

$$\frac{d\Theta}{d\theta} = \frac{d\Theta}{ds}\frac{ds}{d\theta} = -\sin\theta\frac{d\Theta}{ds}$$

$$\Longrightarrow \quad \frac{d^2\Theta}{d\theta^2} = \frac{d^2\Theta}{ds^2}(-\sin\theta)^2 - \frac{d\Theta}{ds}\cos\theta.$$

Plugging the relations above into the $\Theta(\theta)$ equation, and applying the identity of $\sin^2\theta = 1 - \cos^2\theta = 1 - s^2$, we have

$$\frac{d^2\Theta}{ds^2}\sin^2\theta - \frac{d\Theta}{ds}\cos\theta + \frac{\cos\theta}{\sin\theta}\frac{d\Theta}{ds}(-\sin\theta) - \lambda\Theta = 0,$$

which leads to a simplified equation

$$(1-s^2)\frac{d^2\Theta}{ds^2} - 2s\frac{d\Theta}{ds} - \lambda\Theta = 0, \qquad -1 < s < 1. \tag{6.79}$$

The above equation is called a Legendre equation. The Sturm-Liouville eigenvalue problem is a singular one with singularities at $s = \pm 1$, called the north and south pole singularities. From the theory of Legendre equations and polynomials (not discussed here), it has been shown in the literature that

$$\lambda_n = n(n+1), \quad n = 1, 2, \cdots. \tag{6.80}$$

---

[5] Note that the different sign in the right side of the separated equations compared with other problems in previous examples.

The general solutions can be written as Legendre functions

$$\Theta(s) = C_1 P_n(s) + C_2 Q_n(s), \tag{6.81}$$

where $P_n(s)$ is a Legendre polynomial of degree n which is continuous in $[-1, 1]$; $Q_n(s)$ is a second type of solution to the singular Legendre equation which is unbounded at $s = -1$ and $s = 1$. From the differential equation theory and our knowledge on Laplace equations, the solution should be bounded at the center of the sphere ($r = 0$). Thus, we conclude that the coefficient $C_2 = 0$.

Substituting $s = \cos\theta$ into the solution $\Theta(\theta) = P_n(s)$ of the original equation (6.78) we get $\Theta(\theta) = P_n(\cos\theta)$. Since we know the eigenvalues $\lambda_n = n(n+1)$, and $1 + 4\lambda_n = (2n+1)^2$, the solution to the indicial equation for $R(r)$ is

$$s_{n,1} = \frac{-1 - \sqrt{1+4\lambda}}{2} = -n, \qquad s_{n,2} = \frac{-1 + \sqrt{1+4\lambda}}{2} = n.$$

The solution $R_n(r)$ is unbounded at $r = 0$ corresponding to $s_{n,1}$. Thus, we have to have $R_n = r^n$. Therefore, the series solution in the original variable has the following form

$$u(r,\theta) = \sum_{n=1}^{\infty} A_n \left(\frac{r}{a}\right)^n P_n(\cos\theta). \tag{6.82}$$

The way in writing this form is to get A_n in a simple way. The coefficients A_n are determined from the boundary condition $u(a,\theta) = f(\theta)$,

$$u(a,\theta) = \sum_{n=1}^{\infty} A_n P_n(\cos\theta) = f(\theta).$$

To find the coefficients A_n above, we multiply a $P_m(\cos\theta)\sin\theta$ to both sides of the last two terms and integrate

$$\int_0^\pi f(\theta) P_m(\cos\theta) \sin\theta \, d\theta = \sum_{n=1}^{\infty} A_n \int_0^\pi P_n(\cos\theta) P_m(\cos\theta) \sin\theta \, d\theta$$

$$= \sum_{n=1}^{\infty} A_n \int_{-1}^{1} P_n(x) P_m(x) \, dx.$$

From the property of orthogonality of the eigenfunctions and

$$\int_{-1}^{1} P_n^2(x) \, dx = \int_0^\pi P_n^2(\cos\theta) \sin\theta \, d\theta,$$

we obtain

$$A_n = \frac{\displaystyle\int_0^\pi f(\theta) P_n(\cos\theta) \sin\theta \, d\theta}{\displaystyle\int_0^\pi P_n^2(\cos\theta) \sin\theta \, d\theta}. \tag{6.83}$$

Note that the $\sin\theta$ terms in the integrals above correspond to the Jacobian when changing a volume integral from Cartesian coordinates to the spherical ones. Using the property of Legendre polynomials, we can simplify the above expressions further to have,

$$A_n = \frac{2n+1}{2} \int_0^\pi f(\theta) P_n(\cos\theta) \sin\theta \, d\theta. \tag{6.84}$$

We note that the Legendre polynomials can also be expressed as

$$P_n(x) = \frac{1}{2^n n!} \frac{d^n}{dx^n} \left(x^2 - 1\right)^n, \qquad n = 1, 2, \cdots, \tag{6.85}$$

from the orthogonal polynomials theory, see for example [1].

6.9 Special functions related to series solutions of partial differential equations of BVPs

We have seen that with different coordinates, we will have some special Sturm-Liouville eigenvalue problems, often singular, that lead to some special functions if we use the method of separation of variables. We provide a summary in Table 6.1.

Table 6.1. *Some PDEs in different coordinates and related special S-L problems and functions.*

Coordinates	PDE	S-L problems	Eigen-pair
2D polar $0 < r < a$	Laplace eqn. on a disk $\dfrac{\partial^2 u}{\partial r^2} + \dfrac{1}{r}\dfrac{\partial u}{\partial r} = 0$	Euler's equation $r^2 R'' + r R' - n^2 R = 0$	$R(r) = C_n \left(\dfrac{r}{a}\right)^n + \bar{C}_n \left(\dfrac{r}{a}\right)^{-n}$
2D polar $0 < r < a$	Wave eqn. on a disk $\dfrac{\partial u(r,t)}{\partial t} = c^2 \left(\dfrac{\partial^2 u}{\partial r^2} + \dfrac{1}{r}\dfrac{\partial u}{\partial r}\right)$	Bessel's equation $x^2 y'' + x y' + (x^2 - p^2) y = 0$	Bessel's functions $y(x) = C_1 J_p(x) + C_2 Y_p(x)$
3D sphere $0 < r < a$	Laplace eqn. on a sphere $\dfrac{\partial^2 u}{\partial r^2} + \dfrac{2}{r}\dfrac{\partial u}{\partial r} + \dfrac{1}{r^2}\left(\dfrac{\partial^2 u}{\partial \theta^2} + \cot\theta \dfrac{\partial u}{\partial \theta}\right) = 0$	Legendre equation $(-1 < s < 1)$ $(1-s^2)\dfrac{d^2\Theta}{ds^2} - 2s \dfrac{d\Theta}{ds} + \lambda\Theta = 0$	Legendre fun. $(\lambda_n = n(n+1))$ $\Theta(s) = C_1 P_n(s) + C_2 Q_n(s)$

6.10 Exercises

E6.1 Apply the method of separation of variables to solve the 1D wave equation
$$\frac{\partial^2 u}{\partial t^2} = c^2 \frac{\partial^2 u}{\partial x^2}, \quad 0 < x < L \text{ with}$$

$$\frac{\partial u}{\partial x}(0,t) = 0, \quad u(L,t) = 0,$$

$$u(x,0) = f(x), \quad \frac{\partial u}{\partial t}(x,0) = g(x).$$

(a) Let $u(x,t) = X(x)T(t)$, derive the equations for $X(x)$ and $T(t)$.

(b) Solve the related Sturm-Liouville eigenvalue value problem.

(c) Find the series solution to the 1D wave equation.

(d) Apply the derived series solution formula to solve the BVP when $f(x) = |x|$, $g(x) = \sin 4x\pi$, with $L = 2$, $c = 3$. Plot or sketch the series solution, and the partial sums $S_N(x,t)$ (assuming N is large) at $t = 4$.

E6.2 Carry out the method of separation of variables to **solve** the heat equation
$$\frac{\partial u}{\partial t} = c^2 \frac{\partial^2 u}{\partial x^2}, \quad 0 < x < L, \quad u(x,0) = f(x).$$

Also **plot or sketch** the initial $u(x,0) = f(x)$, the series solution, and the partial sums $S_N(x,t)$ assuming N is large enough at $t = 2.5$.

(a) $L = \pi$, $c = 1$, $u(x,0) = 78$, $\frac{\partial u}{\partial x}(0,t) = 0$, $u(L,t) = 0$.

(b) $L = \pi$, $c = 3$, $u(x,0) = 30\sin(10x)$, $u(0,t) = 0$, $u(L,t) = 0$.

(c) $L = \pi$, $c = 2$, $u(0,t) = 0$, $\frac{\partial u}{\partial x}(L,t) = 0$,

$$u(x,0) = \begin{cases} 33x & \text{if } 0 < x \leq \pi/2, \\ 33(\pi - x) & \pi/2 < x < \pi. \end{cases}$$

E6.3 Find the steady state solution (SSS) of the heat equation $\frac{\partial u}{\partial t} = c^2 \frac{\partial^2 u}{\partial x^2}$, $0 < x < L$, $u(x,0) = f(x)$:

(a) $c = 3$, $L = 1$, $u(0,t) = 0$, $u(1,t) = 100$, $f(x) = x^2 \sin x$.

(b) $c = \sqrt{2}$, $L = \pi$, $u(0,t) = 100$, $u(\pi,t) = 100$, $f(x) = \log(1 + x^2)\sin x$.

E6.4 *Solve* the heat equation $\frac{\partial u}{\partial t} = c^2 \frac{\partial^2 u}{\partial x^2}$, $0 < x < L$, $u(x,0) = f(x)$ with non-homogeneous boundary conditions.

(a) $L = 1$, $c = 1$, $f(x) = 30\sin(\pi x)$, $u(0,t) = 100$, $u(1,t) = 0$.

(b) $L = \pi$, $c = 1$, $f(x) = \begin{cases} 33x & \text{if } 0 < x \leq \pi/2, \\ 33(\pi - x) & \pi/2 < x < \pi, \end{cases}$ $u(0,t) = 100$, $u(\pi,t) = 50$.

E6.5 *Solve* the Laplace equation on $[0, 1] \times [0, 2]$ with the following boundary conditions and find its maximum value and minimum values of the solutions.

(a) $u(0,y) = 0$, $u(1,y) = y(2-y)$, $u(x,0) = 0$, $u(x,2) = 0$.

(b) $u(0,y) = 0$, $u(1,y) = 0$, $u(x,0) = \sin(\pi x)$, $u(x,2) = 0$. **Hint:** A normal mode solution.

E6.6 *Solve* the Laplace equation on $[0, a] \times [0, b]$ with the following boundary conditions: $u(0,y) = 0$, $\frac{\partial u}{\partial x}(a,y) = 0$, $u(x,0) = 0$, $u(x,b) = f(x)$.

E6.7 *Solve* the elliptic partial differential equation on $[0, a] \times [0, b]$

$$\frac{\partial^2 u}{\partial x^2} + \alpha^2 \frac{\partial^2 u}{\partial y^2} = 0, \quad (x,y) \in R,$$

$u(x,b) = f_2(x)$, $u(x,0) = 0$, $u(0,y) = 0$, $u(a,y) = 0$.

Hint: Change one of independent variable.

E6.8 Given

$$\frac{\partial u}{\partial t} = u + \frac{\partial^2 u}{\partial x^2}.$$

(a) Classify the following PDE.

(b) Solve the boundary value problem of the PDE on the domain $0 < x < \pi$ with
$$u(0,t) = 0, \quad \frac{\partial u}{\partial x}(\pi,t) = 0, \quad u(x,0) = f(x).$$

(c) Find the steady state solution if it exists.

E6.9 Consider the heat equation.

$$\frac{\partial u}{\partial t} = \frac{\partial^2 u}{\partial x^2} + \frac{\partial^2 u}{\partial y^2}, \quad 0 < x < 2, \quad 0 < y < 1,$$

$u(x,y,0) = \sin(x^2 + y^2)$,

$u(x,0,t) = e^{-t^2}$, $\frac{\partial u}{\partial y}(x,1,t) = 0$, $u(0,y,t) = 0$, $u(2,y,t) = \sin y$.

Write down the PDE and boundary condition(s) of **the steady state solution**.

To solve the steady state solution, do the following:

(a) Let the solution be $u(x,y) = X(x)Y(y)$, derive the ODEs for $X(x)$ and $Y(y)$.

(b) Solve the Sturm-Liouville problem for $Y(y)$. Find the corresponding solution $X(x)$.

(c) Find the series solution for the steady state solution (series solution).

Hint: the solution to the equation $y'' - \lambda^2 y = 0$ can be written as $y(x) = C_1 \cosh(\lambda x) + C_2 \sinh(\lambda x)$.

E6.10 Find the steady state solution of the 2D heat equation,

$$\frac{\partial u}{\partial t} = \Delta u, \qquad x^2 + y^2 < 3,$$

(a): $u(3,\theta,t) = 100 - e^{-t^2}\cos t$, $u(r,\theta,0) = r\sin\theta$. **Hint:** a normal mode.

(b): $u(3,\theta,t) = \begin{cases} 1 & \text{if } 0 < \theta \leq \pi, \\ 0 & \pi < x < 2\pi. \end{cases}$ $\quad u(r,\theta,0) = \log(r^2+1)\cos(7.2\theta)$.

E6.11 The solution to the Laplace equation $\Delta u = 0$ on an annulus $0 < a < r < b$ with the boundary condition $u(a,\theta) = f(\theta)$, $u(b,\theta) = g(\theta)$, $0 < \theta < 2\pi$ can be written as

$$u((r,\theta)) = A_0 + B_0 \log r + \sum_{n=1}^{\infty}\left\{\left(A_n r^n + \frac{B_n}{r^n}\right)\cos n\theta + \left(C_n r^n + \frac{D_n}{r^n}\right)\cos n\theta\right\}.$$

Can you derive the formulas for the coefficients? Also try to solve the problem when $a = 1$, $b = 2$, $f(\theta) = 1$, and $g(\theta) = \sin\theta$.

E6.12 Redo the problem of Example 6.8 with $\frac{\partial u}{\partial r}u(r,0) = 0$, and (a): $u(r,0) = 10$; (b): $u(r,0) = r(1-r)$.

E6.13 Solve the radial symmetric 3D Laplace equation on a sphere with $a = 1$, and (a): $u(a,\theta) = 1$; (b): $u(a,\theta) = \sin\theta\cos\theta$.

E6.14 (Extra Credit): Using the method of separation of variables to solve the following boundary value problem,

$$\frac{\partial^2 u}{\partial t^2} + \alpha u = c^2 \frac{\partial^2 u}{\partial x^2}, \qquad 0 < x < L,$$

$$\frac{\partial u}{\partial x}(0,t) = 0, \qquad u(L,t) = 0,$$

$$u(x,0) = f(x), \qquad \frac{\partial u}{\partial t}(x,0) = g(x),$$

where α is a constant.

Chapter 7

Fourier and Laplace transforms

We have seen the power of various Fourier series in solving boundary value problems of partial differential equations and their applications. If we let L in the Fourier series goto ∞, and replace the summation with integration, then we will have the Fourier transform. The Fourier transform is very useful in terms of theoretical analysis, obtaining analytic solutions of certain PDEs especially those defined in the entire space.

7.1 From the Fourier series to Fourier integral representations

Give a function $f(x) \in L^2(-L, L)$, we already know the Fourier series expansion

$$f(x) = \sum_{n=0}^{\infty} \left\{ a_n \cos \frac{n\pi x}{L} + b_n \sin \frac{n\pi x}{L} \right\}.$$

$$= \sum_{n=0}^{\infty} \left\{ \frac{1}{L} \left(\int_{-L}^{L} f(t) \cos \frac{n\pi t}{L} dt \right) \cos \frac{n\pi x}{L} + \frac{1}{L} \left(\int_{-L}^{L} f(t) \sin \frac{n\pi t}{L} dt \right) \sin \frac{n\pi x}{L} \right\}.$$

Let $\frac{n\pi}{L} = \omega$ or $\frac{1}{L} = \frac{1}{\pi}\frac{w}{n} = \frac{1}{\pi}\Delta\omega$. The expression above becomes

$$f(x) = \sum_{n=0}^{\infty} \left\{ \frac{1}{\pi} \left(\int_{-L}^{L} f(t) \cos \omega t\, dt \right) \cos \omega x + \frac{1}{\pi} \left(\int_{-L}^{L} f(t) \sin \omega t\, dt \right) \sin \omega x \right\} \Delta\omega.$$

Let $L \to \infty$, we get

$$f(x) = \int_0^{\infty} \left(A(\omega) \cos \omega x + B(\omega) \sin \omega x \right) d\omega, \qquad (7.1)$$

where

$$A(\omega) = \frac{1}{\pi} \int_{-\infty}^{\infty} f(t) \cos \omega t \, dt, \quad \text{cosine transform of } f(x); \quad (7.2)$$

$$B(\omega) = \frac{1}{\pi} \int_{-\infty}^{\infty} f(t) \sin \omega t \, dt, \quad \text{sine transform of } f(x). \quad (7.3)$$

The expression (7.1) is called the Fourier integral representation of $f(x)$ which converges to $f(x)$ if $f(x)$ is continuous at a point x; and to $(f(x-) + f(x+))/2$ if $f(x)$ is piecewise continuous.

If we put cosine and sine transforms together and use the trig-identity $\cos(\alpha - \beta) = \cos \alpha \cos \beta + \sin \alpha \sin \beta$, we derive the Fourier transform of $f(x)$ below.

$$f(x) = \frac{1}{\pi} \int_{0}^{\infty} \int_{-\infty}^{\infty} f(t) \left(\cos \omega t \cos \omega x + \sin \omega t \sin \omega x \right) dt \, d\omega$$

$$= \frac{1}{\pi} \int_{0}^{\infty} \int_{-\infty}^{\infty} f(t) \cos(x - t) dt \, d\omega$$

$$= \frac{1}{2\pi} \int_{0}^{\infty} \int_{-\infty}^{\infty} f(t) \left(e^{i\omega(x-t)} + e^{i\omega(x+t)} \right) dt \, d\omega$$

$$= \frac{1}{2\pi} \int_{0}^{\infty} \int_{-\infty}^{\infty} f(t) e^{i\omega(x-t)} dt \, d\omega - \frac{1}{2\pi} \int_{-\infty}^{0} \int_{-\infty}^{\infty} f(t) e^{i\bar{\omega}(x-t)} dt \, d\bar{\omega}$$

$$= \frac{1}{2\pi} \int_{-\infty}^{\infty} \int_{-\infty}^{\infty} f(t) e^{i(\omega - t)} dt \, d\omega$$

$$= \frac{1}{\sqrt{2\pi}} \int_{-\infty}^{\infty} e^{i\omega x} \frac{1}{\sqrt{2\pi}} \int_{-\infty}^{\infty} f(t) e^{-i\omega t} dt \, d\omega$$

$$= \frac{1}{\sqrt{2\pi}} \int_{-\infty}^{\infty} \hat{f}(\omega) e^{i\omega x} d\omega.$$

The Fourier transform of $f(x)$ is defined as

$$\mathcal{F}(f(x)) = \hat{f}(\omega) = \frac{1}{\sqrt{2\pi}} \int_{-\infty}^{\infty} f(x) e^{-i\omega x} dx. \quad (7.4)$$

From the derivation above, we also have.

7.1. From the Fourier series to Fourier integral representations

The Inverse Fourier transform is

$$\mathcal{F}^{-1}(\hat{f}(\omega)) = f(x) = \frac{1}{\sqrt{2\pi}} \int_{-\infty}^{\infty} \hat{f}(\omega) e^{i\omega x} d\omega. \qquad (7.5)$$

Example 7.1. *Find the Fourier transform of $f(x) = e^{-a|x|}$, where $a > 0$ is a constant.*

The Fourier transform can be found directly from the definition.

$$\hat{f}(\omega) = \frac{1}{\sqrt{2\pi}} \int_{-\infty}^{\infty} e^{-a|x|} e^{-i\omega x} dx$$

$$= \frac{1}{\sqrt{2\pi}} \int_{-\infty}^{0} e^{ax} e^{-i\omega x} dx + \frac{1}{\sqrt{2\pi}} \int_{0}^{\infty} e^{-ax} e^{-i\omega x} dx$$

$$= \frac{1}{\sqrt{2\pi}} \int_{0}^{\infty} e^{-ax} e^{i\omega x} dx + \frac{1}{\sqrt{2\pi}} \int_{0}^{\infty} e^{-ax} e^{-i\omega x} dx$$

$$= \frac{1}{\sqrt{2\pi}} \left. \frac{e^{(i\omega - a)x}}{i\omega - a} \right|_{0}^{\infty} + \frac{1}{\sqrt{2\pi}} \left. \frac{-e^{(i\omega + a)x}}{i\omega + a} \right|_{0}^{\infty}$$

$$= \frac{1}{\sqrt{2\pi}} \left(\frac{1}{a + i\omega} + \frac{1}{a - i\omega} \right)$$

$$= \sqrt{\frac{2}{\pi}} \frac{a}{a^2 + \omega^2}.$$

Example 7.2. *Find the Fourier transform of a step function $f(x) = \begin{cases} 1 & \text{if } |x| < a, \\ 0 & \text{if } |x| > a \end{cases}$, where $a > 0$ is a constant.*

The Fourier transform can be found directly from the definition.

$$\hat{f}(\omega) = \frac{1}{\sqrt{2\pi}} \int_{-\infty}^{\infty} f(x) e^{-i\omega x} dx$$

$$= \frac{1}{\sqrt{2\pi}} \int_{-a}^{a} e^{-i\omega x} dx = -\frac{1}{\sqrt{2\pi}} \left. \frac{e^{-i\omega x}}{i\omega} \right|_{-a}^{a}$$

$$= -\frac{1}{\omega i \sqrt{2\pi}} \left(e^{-i\omega a} - e^{i\omega a} \right)$$

$$= -\frac{1}{wi\sqrt{2\pi}}\left(\cos wa - i\sin wa - (\cos wa + i\sin wa)\right)$$

$$= \frac{2}{\sqrt{2\pi}}\frac{\sin wa}{w} = \sqrt{\frac{2}{\pi}}\frac{\sin wa}{w}.$$

Note that $\hat{f}(\omega)$ has a removable singularity at $\omega = 0$ since

$$\hat{f}(0) = \lim_{\omega \to 0} \sqrt{\frac{2}{\pi}}\frac{\sin wa}{\omega} = a\sqrt{\frac{2}{\pi}}.$$

Example 7.3. *Find the Fourier transform of a point source function $f(x) = \delta(x)$, a special function defined only in the sense of distribution below,*

$$\int f(x)\delta(x-\alpha)dx = f(\alpha) \tag{7.6}$$

if α is in the domain of the integration.

The Fourier transform can be found directly from the definition,

$$\hat{f}(\omega) = \frac{1}{\sqrt{2\pi}}\int_{-\infty}^{\infty}\delta(x)e^{-i\omega x}dx = \frac{1}{\sqrt{2\pi}}.$$

Note that the point source function $\delta(x)$ is called a Dirac delta function, which can be regarded as a 'limit' of the following non-negative function whose graph about the x-axis has a unit area

$$\delta_\epsilon(x) = \begin{cases} \dfrac{1-|x|}{\epsilon} & \text{if } |x| \leq \epsilon, \\ 0 & \text{Otherwise.} \end{cases} \tag{7.7}$$

It is easy to show that

$$\lim_{\epsilon \to 0}\int f(x)\delta_\epsilon(x-\alpha)dx = f(\alpha).$$

Such a $\delta_\epsilon(x)$ is not unique, for example, the following function plays the same role

$$\delta_\epsilon(x) = \begin{cases} \dfrac{1}{4\epsilon}\left(1+\cos\dfrac{\pi x}{2\epsilon}\right) & \text{if } |x| \leq 2\epsilon, \\ 0 & \text{Otherwise.} \end{cases} \tag{7.8}$$

The Dirac delta function can be regarded as the 'weak derivative' of the Heaviside function

$$H(x) = \begin{cases} 1 & \text{if } x > 0, \\ 0 & \text{Otherwise.} \end{cases} \tag{7.9}$$

7.1.1 Properties of the Fourier transform

It is obvious that if α and β are two complex or real constants, then

$$\mathcal{F}\left(\alpha f(x) + \beta g(x)\right) = \alpha \mathcal{F}(f(x)) + \beta \mathcal{F}(g(x)). \tag{7.10}$$

Theorem 7.1. *Let $\hat{u}$ be the Fourier transform of a function $u \in L^2$, then*

$$\widehat{\frac{\partial u}{\partial x}} = i\omega \hat{u}, \tag{7.11}$$

$$\widehat{\frac{\partial \hat{u}}{\partial \omega}} = -ixu, \tag{7.12}$$

$$\hat{\hat{u}} = u. \tag{7.13}$$

Proof: From the definition, we have

$$\hat{\hat{u}} = \frac{1}{\sqrt{2\pi}} \int_{-\infty}^{\infty} e^{i\omega x} \hat{u}\, d\omega = u. \tag{7.14}$$

For the partial derivatives $\frac{\partial u}{\partial x}$, first from the definition, we have

$$\frac{\partial u}{\partial x} = \frac{1}{\sqrt{2\pi}} \int_{-\infty}^{\infty} \widehat{\frac{\partial u}{\partial x}} e^{i\omega x} d\omega.$$

On the other hand, if we take the partial derivative of (7.14) with respect to x assuming that we can switch the integration and the partial derivative, we get

$$\frac{\partial u}{\partial x} = \frac{1}{\sqrt{2\pi}} \int_{-\infty}^{\infty} \frac{\partial}{\partial x} \left(e^{i\omega x} \hat{u}(\omega)\right) d\omega$$

$$= \frac{1}{\sqrt{2\pi}} \int_{-\infty}^{\infty} i\omega e^{i\omega x} \hat{u}(\omega)\, d\omega.$$

The inside expressions should be the same, that is, $\widehat{\frac{\partial u}{\partial x}} = i\omega \hat{u}$.

If we switch the position between ω and x, u and $\hat{u}$ in the expression above, we get

$$\frac{\partial \hat{u}}{\partial \omega} = \frac{1}{\sqrt{2\pi}} \int_{-\infty}^{\infty} e^{-i\omega x} \widehat{\frac{\partial \hat{u}}{\partial \omega}} dx,$$

and by differentiating the Fourier transform

$$\hat{u}(\omega) = \frac{1}{\sqrt{2\pi}} \int_{-\infty}^{\infty} u(x) e^{-i\omega x} u\, dx$$

with respect to ω we get

$$\frac{\partial \hat{u}}{\partial \omega} = \frac{1}{\sqrt{2\pi}} \int_{-\infty}^{\infty} -ix e^{-i\omega x} u \, dx.$$

Thus we get $\widehat{\frac{\partial \hat{u}}{\partial \omega}} = -ixu$, which completes the proof.

It is easy to generalize the equality to high order derivatives to get,

$$\widehat{\frac{\partial^m u}{\partial x^m}} = (i\omega)^m \hat{u} \tag{7.15}$$

i.e., we can remove derivatives using the Fourier transform.

Parseval's relation: Under the Fourier transform, we have $\|\hat{u}\|_2 = \|u\|_2$ or

$$\int_{-\infty}^{\infty} |\hat{u}|^2 d\omega = \int_{-\infty}^{\infty} |u|^2 dx. \tag{7.16}$$

7.1.2 The convolution theorem of the Fourier transform

For some functions, it may be easier to obtain the inverse Fourier transform using the convolution theorem:

$$\mathcal{F}^{-1}[F(\omega)G(\omega)] = f * g = \int_{-\infty}^{\infty} f(y)g(x-y)dy, \tag{7.17}$$

where $f * g$ is called the *convolution* of f and g. Thus, we also have

$$\mathcal{F}\left[\int_{-\infty}^{\infty} f(y)g(x-y)dy\right] = F(\omega)G(\omega). \tag{7.18}$$

We show an example below about an application of the convolution theorem. More examples can be found in the area of signal processing.

Example 7.4. *Solve for $y(x)$ from the integral equation,*

$$\int_{-\infty}^{\infty} \frac{y(\omega) d\omega}{(x-\omega)^2 + a^2} = \frac{1}{x^2 + b^2}$$

assuming that $b > a > 0$.

Solution: The left hand side looks like a convolution. Note that

$$\int_{-\infty}^{\infty} \frac{e^{-i\omega x}}{x^2 + b^2} dx = \int_{-\infty}^{0} \frac{e^{-i\omega x}}{x^2 + b^2} dx + \int_{0}^{\infty} \frac{e^{-i\omega x}}{x^2 + b^2} dx$$

$$= \int_{0}^{\infty} \frac{e^{i\omega x} + e^{-i\omega x}}{x^2 + b^2} dx = \frac{\pi}{b} e^{-\omega b}.$$

We take the Fourier transform of the integral equation on both sides, that is,

$$\mathcal{F}\{y(x)\}\,\mathcal{F}\left\{\frac{1}{x^2+b^2}\right\} = \mathcal{F}\left\{\frac{1}{x^2+b^2}\right\}$$

and apply the convolution theorem to get

$$Y(\omega)\frac{\pi}{a}e^{-\omega a} = \frac{\pi}{b}e^{-\omega b} \implies Y(\omega) = \frac{a}{b}e^{-(b-a)\omega}.$$

After the inverse Fourier transform, we have

$$y(x) = \frac{(b-a)a}{b\pi\left(x^2+(b-a)^2\right)}.$$

7.2 Use the Fourier transform to solve PDEs

The Fourier transform is a powerful tool to solve some partial differential equations, particularly for some Cauchy problems as illustrated below.

Example 7.5. *Consider the Cauchy problem below,*

$$u_t + au_x = 0, \quad -\infty < x < \infty, \quad t > 0, \quad u(x,0) = u_0(x),$$

which is an advection equation, or a one-way wave equation. This is a Cauchy problem since the spatial variable is defined in the entire space and $t \geq 0$. Applying the Fourier transform to the equation and the initial condition, we get

$$\widehat{u_t} + a\widehat{u_x} = 0, \quad \text{or} \quad \hat{u}_t + ai\omega\hat{u} = 0, \quad \hat{u}(\omega,0) = \hat{u}_0(\omega),$$

which is an initial value problem of an ordinary differential equation. The solution is

$$\hat{u}(\omega,t) = \hat{u}(\omega,0)\,e^{-ia\omega t} = \hat{u}_0(\omega)\,e^{-ia\omega t}$$

for $\hat{u}(\omega)$. The solution to the original advection equation is obtained from the inverse Fourier transform,

$$u(x,t) = \frac{1}{\sqrt{2\pi}}\int_{-\infty}^{\infty} e^{i\omega x}\hat{u}_0(\omega)\,e^{-ia\omega t}\,d\omega$$

$$= \frac{1}{\sqrt{2\pi}}\int_{-\infty}^{\infty} e^{i\omega(x-at)}\hat{u}_0(\omega)\,d\omega$$

$$= u(x-at,0) = u_0(x-at)$$

which is the same as that in Chapter 2. It is noted that the solution to the advection equation does not change the shape of the initial condition, but simply propagates along the characteristic line $x - at = 0$, and the Parseval's identity,

$$\|u\|_2 = \|\hat{u}\|_2 = \|\hat{u}(\omega,0)e^{-ia\omega t}\|_2 = \|\hat{u}(\omega,0)\|_2 = \|u_0\|_2 \,.$$

Example 7.6. *Consider*

$$u_t = \beta u_{xx}, \quad -\infty < x < \infty, \quad t > 0, \quad u(x,0) = u_0(x), \quad \lim_{|x|\to\infty} u = 0,$$

which is a heat (or diffusion) equation. Once again applying the Fourier transform to the PDE and the initial condition, we obtain

$$\widehat{u_t} = \widehat{\beta u_{xx}}, \quad \text{or} \quad \hat{u}_t = \beta(i\omega)^2 \hat{u} = -\beta\omega^2 \hat{u}, \quad \hat{u}(\omega,0) = \hat{u}_0(\omega).$$

The solution of this ODE is

$$\hat{u}(\omega,t) = \hat{u}(\omega,0)\, e^{-\beta\omega^2 t}\,.$$

Consequently, if $\beta > 0$, from the Parseval's relation, we have

$$\|u\|_2 = \|\hat{u}\|_2 = \|\hat{u}(\omega,0)e^{-\beta\omega^2 t}\|_2 \le \|u_0\|_2\,.$$

Actually, it can be seen that $\lim_{t\to\infty} \|u\|_2 = 0$. That is why the second order partial derivative term is called a diffusion or dissipation term. The L^2 norm is often regarded as an energy in some physical applications. In a heat equation, the energy is decreasing with the time. If $\beta < 0$, then $\lim_{t\to\infty} \|u\|_2 = \infty$, the partial differential equation is dynamically unstable. The partial differential equation is called a backward heat equation, which has application in financial mathematics with terminal (backward) boundary conditions.

Example 7.7. *Dispersive waves. Consider*

$$u_t = \frac{\partial^{2m+1} u}{\partial x^{2m+1}} + \frac{\partial^{2m} u}{\partial x^{2m}} + l.o.t.,$$

where m is a non-negative integer. For the simplest case $u_t = u_{xxx}$, we have

$$\widehat{u_t} = \widehat{\beta u_{xxx}}, \quad \text{or} \quad \hat{u}_t = \beta(i\omega)^3 \hat{u} = -i\omega^3 \hat{u}\,,$$

and the solution of this initial value problem of the ODE above is

$$\hat{u}(\omega,t) = \hat{u}(\omega,0)\, e^{-i\omega^3 t}\,.$$

Therefore
$$\|u\|_2 = \|\hat{u}\|_2 = \|\hat{u}(\omega,0)\|_2 = \|u(\omega,0)\|_2,$$
and the solution to the original PDE can be expressed as
$$u(x,t) = \frac{1}{\sqrt{2\pi}} \int_{-\infty}^{\infty} e^{i\omega x} \hat{u}_0(\omega) e^{-i\omega^3 t} d\omega$$
$$= \frac{1}{\sqrt{2\pi}} \int_{-\infty}^{\infty} e^{i\omega(x-\omega^2 t)} \hat{u}_0(\omega) d\omega.$$

Evidently, the Fourier component with wave number ω propagates with velocity ω^2, so waves mutually interact but there is no diffusion.

Example 7.8. *PDEs with higher order derivatives. Consider*
$$u_t = \alpha \frac{\partial^{2m} u}{\partial x^{2m}} + \frac{\partial^{2m-1} u}{\partial x^{2m-1}} + l.o.t.,$$
where m is a non-negative integer. The Fourier transform yields
$$\hat{u}_t = \alpha(i\omega)^{2m}\hat{u} + \cdots = \begin{cases} -\alpha\omega^{2m}\hat{u} + \cdots & \text{if } m = 2k+1, \\ \alpha\omega^{2m}\hat{u} + \cdots & \text{if } m = 2k, \end{cases}$$
hence
$$\hat{u} = \begin{cases} \hat{u}(\omega,0) e^{-\alpha i \omega^{2m} t} + \cdots & \text{if } m = 2k+1, \\ \hat{u}(\omega,0) e^{\alpha i \omega^{2m} t} + \cdots & \text{if } m = 2k. \end{cases}$$

From the above relations, we can know whether they partial differential equations are dynamically stable or not. For example, $u_t = u_{xx}$ and $u_t = -u_{xxxx}$ are dynamically stable, whereas $u_t = -u_{xx}$ and $u_t = u_{xxxx}$ are dynamically unstable.

7.3 The Laplace transform

The Fourier transform is for functions that are defined in the entire space $(-\infty, \infty)$ while the Laplace transform is for functions that are defined in half space $(0, \infty)$ such as time variable $t > 0$. The Laplace transform for a function $f(t)$ is defined as
$$\mathcal{L}(f)(s) = F(s) = \int_0^{\infty} f(t) e^{-st} dt, \quad (7.19)$$
where s is in the complex number set. A necessary condition for the existence of the integral is that f must be locally integrable on $[0, \infty)$. The Laplace transformation from the time domain to the frequency, also referred as s-domain, transforms

ordinary differential equations to algebraic equations and convolutions to multiplications.

Example 7.9. Find the Laplace transform of $f(t) = 1$, $f(t) = t$, $f(t) = e^{\alpha t}$, and $f(t) = \sin(\omega t)$.

We apply the Laplace transform formula to get:

$$\mathcal{L}(1)(s) = \int_0^\infty 1 \cdot e^{-st} \, dt = -\frac{1}{s} e^{-st} \Big|_{t=0}^\infty = \frac{1}{s}.$$

$$\mathcal{L}(t)(s) = \int_0^\infty t \cdot e^{-st} \, dt = \left(-\frac{t}{s} e^{-st} - \frac{e^{-st}}{s^2} \right) \Big|_{t=0}^\infty = \frac{1}{s^2}.$$

$$\mathcal{L}(e^{\alpha t})(s) = \int_0^\infty e^{\alpha t} \cdot e^{-st} \, dt = -\frac{1}{s-a} e^{-(s-a)t} \Big|_{t=0}^\infty = \frac{1}{s-a},$$

$$\mathcal{L}(\sin(\omega t))(s) = \int_0^\infty \sin(\omega t) \cdot e^{-st} \, dt = \frac{\omega}{\omega^2 + s^2}.$$

The last one is from the integral table: $\int \sin(\omega t) \cdot e^{-st} dt = -e^{-st} \frac{s \sin(\omega t) + \omega \cos(\omega t)}{s^2 + \omega^2}$.

One of the most important properties of Laplace transform is that we can get rid of one derivatives from the following identities:

$$\mathcal{L}(f') = s\mathcal{L}(f) - f(0), \tag{7.20}$$

$$\mathcal{L}(f^{(n)}) = s^n \mathcal{L}(f) - s^{n-1} f(0) - s^{n-2} f'(0) - \cdots - f^{(n-1)}(0). \tag{7.21}$$

Proof: If we repeatedly apply the integration by parts, then we have

$$\mathcal{L}(f^{(n)}(t))(s) = \int_0^\infty f^{(n)}(t) \cdot e^{-st} \, dt = f^{(n-1)}(t) \cdot e^{-st} \Big|_{t=0}^\infty + \int_0^\infty s f^{(n-1)}(t) \cdot e^{-st} \, dt$$

$$= -f^{(n-1)}(0) + f^{(n-2)}(t) \cdot se^{-st} \Big|_{t=0}^\infty + \int_0^\infty s f^{(n-3)}(t) \cdot s^2 e^{-st} \, dt$$

$$= -f^{(n-1)}(0) - sf^{(n-2)}(0) + \cdots + \int_0^\infty s^n f(t) \cdot e^{-st}$$

$$= -f^{(n-1)}(0) - sf^{(n-2)}(0) - \cdots s^{n-1} f(0) + s^n \mathcal{L}(f)(s).$$

The first identity is the directly application of the above.

7.3.1 The inverse Laplace transform and the convolution theorem

The definition of the inverse Laplace transform is quite technical and involves integrations on the complex plane. Intuitively, if the Laplace transform a function $f(t)$ is $F(s)$, i.e. $\mathcal{L}(f)(s) = F(s)$, then $f(t)$ is called an *inverse Laplace transform* of $F(s)$. We write symbolically $\mathcal{L}^{-1}(F)(s) = f(t)$. Technically, the inverse Laplace transform can be expressed as

$$\mathcal{L}^{-1}(F)(s) = f(t) = \frac{1}{2\pi i} \lim_{T \to \infty} \int_{\gamma - iT}^{\gamma + iT} F(s) e^{st} \, dt. \tag{7.22}$$

where the integration is done along the vertical line $Re(s) = \gamma$ in the complex plane. In practice, computing the complex integral can be done by the Cauchy residue theorem. Note that not all functions have inverse Laplace transform.

The inverse Laplace formula is not very useful due to its complexity. Fortunately, most of useful and practical inverse Laplace transforms can be found in the literature and on the Internet. For some functions, it is easier to obtain the inverse Laplace formula using the convolution theorem:

$$\mathcal{L}^{-1}[F(s)G(s)] = f * g = \int_0^t f(y) g(t-y) dy, \tag{7.23}$$

where $f * g$ is called the *convolution* of f and g. Thus, we also have

$$\mathcal{L}\left[\int_0^t f(y) g(t-y) dy\right] = F(s) G(s). \tag{7.24}$$

Example 7.10. *Find the inverse Laplace transform of* $\dfrac{2}{s^2(s^2+4)}$.

Solution: We set $F(s) = \dfrac{2}{s^2}$, thus $f(t) = 2t$, and $G(s) = \dfrac{1}{s^2+4}$, thus $g(t) = \dfrac{\sin(2t)}{2}$. The convolution of $f * g$ is

$$\int_0^t 2(t-y) \sin(2y) dy = \frac{t}{2} - \frac{\sin(2t)}{4}.$$

Thus, we conclude

$$\mathcal{L}^{-1}[F(s)G(s)] = \mathcal{L}^{-1}\left[\frac{2}{s^2(s^2+4)}\right] = \int_0^t 2(t-y) \sin(2y) dy = \frac{t}{2} - \frac{\sin(2t)}{4},$$

which is the inverse Laplace transform.

Similar to the Fourier transform, the Laplace transform can be used to solve differential equations by eliminating derivatives of one variable. For a linear ordinary differential equation, the Laplace transform reduces the ODE to an algebraic equation, which can then be solved by algebra. The original differential equation can then be solved by applying the inverse Laplace transform.

Example 7.11. *Use the Laplace transform to solve the initial value problem:*

$$y''(t) + y(t) = 2, \quad y(0) = 0, \quad y'(0) = 1.$$

Solution: We apply the Laplace transform to the ODE $\mathcal{L}[y'' + y] = \mathcal{L}(2)$, and apply property for derivatives to obtain

$$s^2 Y(s) - sy(0) - y'(0) + Y = \frac{2}{s}.$$

Apply the initial conditions, we get an algebraic equation for $Y(s)$

$$\left(s^2 + 1\right) Y(s) - 1 = \frac{2}{s}, \quad \Longrightarrow$$

$$Y(s) = \frac{1}{s^2 + 1} + \frac{2}{s(s^2 + 1)} = \frac{1}{s^2 + 1} + \frac{2}{s} - \frac{2s}{s^2 + 1}.$$

By looking at a mathematical handbook for the Laplace transform we get the solution $y(t) = \sin t + 2 - 2\cos t$. It is easy to check that $y(t)$ satisfies the ODE and the initial conditions.

7.4 Exercises

E7.1 Show that if $f(x)$ is an even function, then its Fourier transform can be expressed as the cosine transform

$$\mathcal{F} = \frac{1}{\sqrt{2\pi}} \int_0^\infty f(\omega) \cos \omega x \, dx.$$

Similarly, if $f(x)$ is an odd function, find the similar relation in terms of the sine transform

$$\mathcal{F} = \frac{1}{\sqrt{2\pi}} \int_0^\infty f(\omega) \sin \omega x \, dx.$$

7.4. Exercises

E7.2 (a): Find the Fourier transform of a pulse defined as:

$$P(x) = \begin{cases} 1 & \text{if } -\frac{1}{2} \leq x \leq \frac{1}{2}, \\ 0 & \text{otherwise}; \end{cases}$$

and a triangular pulse, defined as:

$$T(x) = \begin{cases} 1 - |t| & \text{if } -1 \leq x \leq 1, \\ 0 & \text{otherwise}; \end{cases}$$

(b): Show that $T(x) = P * P$, That is, the convolution of $P(x)$ and itself.

(c): Use the convolution theorem to find the Fourier transform

$$\mathcal{F}\{T\} = \mathcal{F}\{P * P\} = \mathcal{F}\{P\} \cdot \mathcal{F}\{P\} = C \frac{\sin^2(\pi\omega)}{\omega^2},$$

Find the constant C.

E7.3 Using the convolution theorem to show that the solution to the integral equation

$$y(x) = g(x) + \int_{-\infty}^{\infty} y(t) r(x-t) dt$$

is

$$y(x) = \frac{1}{\sqrt{2\pi}} \int_{-\infty}^{\infty} \left(\frac{G(\omega)}{1 - R(\omega)} \right) e^{i\omega x} d\omega,$$

where $G(\omega)$ and $R(\omega)$ are the Fourier transform of $g(x)$ and $r(x)$, respectively.

E7.4 Find the Laplace transformation of the following functions.

(a) $f(t) = C$.
(b) $f(t) = \cos(\omega t)$.
(c) $f(t) = \sinh(t)$.
(d) $f(t) = \cosh(t)$.

E7.5 Find the inverse Laplace transformation of the following

(a). $F(s) = \dfrac{1}{s^3}$.

(b). $F(s) = \dfrac{1}{s-a}$.

(c). $F(s) = \dfrac{1}{s^2 + a^2}$.

Chapter 8

Numerical solution techniques

Analytic solutions techniques are important for solving and understanding differential equations. We should try our best effort to get analytic solutions so that we can analyze and understand the solution behaviors. Unfortunately, many problems are difficult, if not impossible to find analytic solutions, particularly for partial differential equations. Thus, we need to find different ways to solve and analyze differential equations. Series solution techniques have been shown to be effective in solving and analyzing differential equations. Nevertheless, Series solution techniques can be difficult for high dimensional problems and have limitations on boundary conditions.

The rapid development in modern computers has provided another powerful tool in solving differential equations, which is called numerical solutions of differential equations. Nowadays, many applications such as weather forecasts, space shuttles lunches, robots, heavily depend on super-computer simulations. There are tons of books, software packages, numerical methods, online classes for solving differential equations. It is totally a new area of study and research. Here, we just introduce a few examples so that interested readers can get a glance of the powerful tool and can pursue further if needed. We can see that for some problems, numerical approaches maybe much simpler than series or other analytical solution techniques. Note also that numerical methods can also help theoretical study of differential equations.

There are many different numerical methods that can be applied to solve differential equations, for examples, finite different methods, finite element methods, finite volume methods. Usually different methods have advantages/limitations compared with other ones for solving differential equations and applications.

While Maple is a symbolic package, Matlab is a multi-paradigm numerical computing environment and proprietary programming language developed by MathWorks. One can call Maple directly from the Matlab environment. Matlab has a variety of numerical methods built in such as linear algebra, numerical approximations, spline library, and many toolboxes for solving various mathematical problems through computing. Note that, Matlab is a simple and easy tool to use. In some sense, it is more like a super calculator. Nevertheless, it is not the most efficient way for large scale simulations and super-computing.

For ordinary differential equations or systems of ordinary differential equations with prescribed initial conditions, particularly, the first order system of ODEs with an initial condition,

$$\frac{d\mathbf{y}}{dt} = \mathbf{F}(t, \mathbf{y}), \qquad \mathbf{y}(t_0) = \mathbf{y}_0, \tag{8.1}$$

where $\mathbf{y}_0$ is a known initial condition, we can use the Matlab ODE Suite toolbox to solve the problem numerically, see for example, [11]. The toolbox is powerful and often enough for many practical applications. For a high order ordinary differential equation, often it is easier to convert the ODE to a first order system as the above standard form.

In this chapter, we briefly explain finite difference methods (FDM) for some differential equations of boundary value problems. We refer the interested reader to [7, 8, 10, 14, 15] for introductions on this topic. In a finite difference method, instead of finding solution everywhere, we seek approximate solutions at a finite number of points, called grid points. The second aspect of a finite difference method is to approximate derivatives using function values at grid points so that a differential equation becomes an algebraic system of equations. Some commonly used finite difference formulas are listed below

$$u'(x) = \lim_{h \to 0} \frac{u(x+h) - u(x)}{h} \approx \frac{u(x+h) - u(x)}{h}, \tag{8.2}$$

$$u'(x) = \lim_{h \to 0} \frac{u(x+h) - u(x-h)}{2h} \approx \frac{u(x+h) - u(x-h)}{2h}, \tag{8.3}$$

$$u''(x) = \lim_{h \to 0} \frac{u(x-h) - 2u(x) + u(x+h)}{h^2} \approx \frac{u(x+h) - 2u(x) + u(x+h)}{h^2} \tag{8.4}$$

if h is small enough.

The simplest method for solving (8.1) may be the forward Euler's method that uses a time marching approach to obtain an approximate solution at the time interval Δt starting from the initial condition as

$$\frac{\mathbf{y}^{n+1} - \mathbf{y}^n}{\Delta t} = \mathbf{f}(t^n, \mathbf{y}^n), \qquad n = 0, 1, \cdots. \tag{8.5}$$

8.1. Finite difference methods for two-point boundary value problems

The method is first order accurate, that is, $\|\mathbf{y}^n - \mathbf{y}(t^n, \mathbf{y}(t^n))\| \leq C\Delta t$. The scheme is conditionally stable meaning that we can not take very large Δt. We can give a reasonable guess of $\Delta t < 1$. Theoretically, the method is stable if we choose Δt such that $|\Delta t \lambda_i \left(\frac{D\mathbf{f}}{D\mathbf{y}}(t_0, \mathbf{y}(t_0))\right)| \leq 1$ for all i's, where $\lambda_i \left(\frac{D\mathbf{f}}{D\mathbf{y}}(t_0, \mathbf{y}(t_0))\right)$ are the eigenvalues of the Jacobi matrix of $\mathbf{f}$ at $(t_0, \mathbf{y}(t_0))$.

It is quite easy to use the Matlab ODE Suite to solve a system of first order ODEs of an initial value problem and visualize the results. The Matlab ODE Suite is a collection of five user friendly finite difference codes for solving initial value problems given by first-order systems of ordinary differential equations and plotting their numerical solutions. The three codes ode23, ode45, and ode113 are designed to solve non-stiff problems and the two codes ode23s and ode15s are designed to solve both stiff and non-stiff problems. The mathematical and software developments and analysis are given in [11]. The Matlab ODE Suite is based on Runge-Kutta methods and can choose time step size adaptively.

As a demonstration, we solve the non-dimensionalized Lotka-Volterra predator-prey model of the following system,

$$y_1' = y_1 - y_1 y_2,$$
$$y_2' = -ay_2 + y_1 y_2, \qquad (8.6)$$
$$y_1(0) = p_1, \qquad y_2(0) = p_2,$$

where p_1 and p_2 are two constant, $y_1(t)$ is the population of a prey while $y_1(t)$ is the population of a predator. Under certain conditions, predator and prey can co-exit.

We define the system in a Matlab function called *prey_prd.m* whose contents are

```
function yp = prey_prd(t,y)
global a
k = length(y); yp = zeros(k,1);
yp(1) = y(1) - y(1)*y(2) ;
yp(2) = -a*y(2) + y(1)*y(2);
```

To solve the problem, we write a Matlab script file called *prey_prd_drive.m* whose contents are the following.

```
global a
a = 0.5; t0 = 0; y0 = [0.01 0.01];  tfinal=200;
[t y] = ode23s('prey_prd',[t0,tfinal],y0);
```

```
y1 = y(:,1); y2=y(:,2);              % Extract solution components.
figure(1); subplot(211); plot(t,y1); title('Population of prey')
subplot(212); plot(t,y2);            title('Population of predator')
figure(2); plot(y1,y2)               % Phase plot
xlabel('prey'); ylabel('predator'); title('phase plot')
```

In Figure 8.1, we plot the computed solution for the parameters $a = 0.5$, the initial data is $y_1(0) = y_2(0) = 0.01$. The final time is $T = 200$. The left plot are the solution of each component against time. We can observe that the solution changes rapidly in some regions indicating the stiffness of the problem. The right plot is the phase plot, that is, the plot of one component against the other. The phase plot is more like a closed curve in the long run indicating the existence of the limiting cycle of the model.

(a) (b)

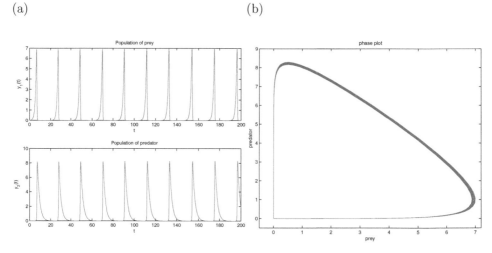

Figure 8.1. *Plots of the solution of the prey-predator model from $t = 0$ to $t = 200$ in which we can see the prey and predator co-exist. (a), solution plots against time; (b), the phase plot in which limit cycle can be seen.*

8.1 Finite difference methods for two-point boundary value problems

We start with a one-dimensional Sturm-Liouville problem of the following,

$$u''(x) = f(x), \quad 0 < x < L, \quad u(0) = u_a, \quad u(1) = u_b,$$

8.1. Finite difference methods for two-point boundary value problems

to illustrate the general procedure using a finite difference method as follows. Note that when $u_a = 0$, $u_b = 0$, and $f(x) = \lambda u$, we have a Sturm-Liouville eigenvalue problem.

8.1.1 Outline of a finite difference method for a two-point BVP

1 <u>Generate a grid.</u> A grid, or a mesh, is a finite set of points on which we seek an approximate solution to the differential equation. For example, we can use a uniform Cartesian grid

$$x_i = i\,h, \quad i = 0, 1, \cdots n, \quad h = \frac{1}{n}.$$

2 <u>Represent the derivative by a finite difference formula</u> at every grid point where the solution is unknown, to get an algebraic system of equations. At each grid point x_i we replace the differential equation in the model problem by

$$\frac{u(x_i - h) - 2u(x_i) + u(x_i + h)}{h^2} = f(x_i) + error,$$

where the error is called the *local truncation error*.

Thus, we obtain the finite difference solution (an approximation) to $u(x)$ at all x_i as the solution $U_i \approx u(x_i)$ (if it exists) of the following linear system of algebraic equations:

$$\frac{u_a - 2U_1 + U_2}{h^2} = f(x_1)$$

$$\frac{U_1 - 2U_2 + U_3}{h^2} = f(x_2)$$

$$\cdots \cdots = \cdots$$

$$\frac{U_{i-1} - 2U_i + U_{i+1}}{h^2} = f(x_i)$$

$$\cdots \cdots = \cdots$$

$$\frac{U_{n-3} - 2U_{n-2} + U_{n-1}}{h^2} = f(x_{n-2})$$

$$\frac{U_{n-2} - 2U_{n-1} + u_b}{h^2} = f(x_{n-1}).$$

Note that the finite difference approximation at each grid point involves solution values at three grid points, *i.e.*, at x_{i-1}, x_i, and x_{i+1}. The set of these three grid points is called the *finite difference stencil*.

3 Solve the system of algebraic equations to get an approximate solution at each grid point. The system of algebraic equations can be written in the matrix and vector form ($Ax = b$),

$$\begin{bmatrix} -\frac{2}{h^2} & \frac{1}{h^2} & & & & \\ \frac{1}{h^2} & -\frac{2}{h^2} & \frac{1}{h^2} & & & \\ & \frac{1}{h^2} & -\frac{2}{h^2} & \frac{1}{h^2} & & \\ & & \ddots & \ddots & \ddots & \\ & & & \frac{1}{h^2} & -\frac{2}{h^2} & \frac{1}{h^2} \\ & & & & \frac{1}{h^2} & -\frac{2}{h^2} \end{bmatrix} \begin{bmatrix} U_1 \\ U_2 \\ U_3 \\ \vdots \\ U_{n-2} \\ U_{n-1} \end{bmatrix} = \begin{bmatrix} f(x_1) - u_a/h^2 \\ f(x_2) \\ f(x_3) \\ \vdots \\ f(x_{n-2}) \\ f(x_{n-1}) - u_b/h^2 \end{bmatrix} \quad (8.7)$$

Note that it can be shown that the coefficient matrix is a symmetric negative definite matrix and it is invertible. There are various computer packages designed to solve such a system of equations.

4 Implement and debug the computer code. Run the program to get the output. Analyze and visualize the results (tables, plots etc.).

5 Estimate errors. We can show that the finite difference method is consistency and stability, which implies the convergence of the finite difference method. In fact, we can show the convergence is pointwise, i.e.,

$$\lim_{h \to 0} \|u(x_i) - U_i\|_\infty = 0. \quad (8.8)$$

8.1.2 A Matlab code for the model problem

Below is a Matlab function called *two_point.m* for the model problem. We use this Matlab function to illustrate how to convert the algorithm to a computer code.

```
function [x,U] = two_point(a,b,ua,ub,f,n)
```

```
%%%%%%%%%%%%%%%%%%%%%%%%%%%%%%%%%%%%%%%%%%%%%%%%%%%%%%%%%%%%%%%%%%%%%
% This matlab function two_point solves the following two-point     %
% boundary value problem: u''(x) = f(x) using the centered finite   %
% difference scheme.                                                %
% Input:                                                            %
```

8.1. Finite difference methods for two-point boundary value problems 153

```
%   a, b: Two end points.                                              %
%   ua, ub: Dirichlet boundary conditions at a and b.                  %
%   f: external function f(x).                                         %
%   n: number of grid points.                                          %
% Output:                                                              %
%   x: x(1),x(2),...x(n-1) are grid points                             %
%   U: U(1),U(2),...U(n-1) are approximate solution at grid points     %
%%%%%%%%%%%%%%%%%%%%%%%%%%%%%%%%%%%%%%%%%%%%%%%%%%%%%%%%%%%%%%%%%%%%%%%%

h = (b-a)/n; h1=h*h;

A = sparse(n-1,n-1);
F = zeros(n-1,1);

for i=1:n-2,
  A(i,i) = -2/h1; A(i+1,i) = 1/h1; A(i,i+1)= 1/h1;
end
  A(n-1,n-1) = -2/h1;

for i=1:n-1,
  x(i) = a+i*h;
  F(i) = feval(f,x(i));
end
  F(1) = F(1) - ua/h1;
  F(n-1) = F(n-1) - ub/h1;

U = A\F;

return
%%%%--------- End of the program --------------------------------
```

We can call the Matlab function two_point directly in a Matlab command window. A better way is to put all Matlab commands in a Matlab script file (called an M-file), referred to as *main.m* here. The advantage of using a script file is to keep a record, and can also be re-visited or modified whenever we want.

To illustrate, suppose the interval of integration is $[0, 1]$, $f(x) = -\pi^2 \cos(\pi x)$, $u(0) = 0$ and $u(1) = -1$. A sample Matlab M-file is then as follows.

```
%%%%%%% Clear all unwanted variables and graphs.
  clear; close all
%%%%%%% Input


a=0; b=1; n=40;
ua=1; ub=-1;


%%%%% Call the solver: U is the FD solution at the grid points.

[x,U] = two_point(a,b,ua,ub,'f',n);


%%%%%%%%%%%%%%% Plot and show the error %%%%%%%%%%%%%%%%


plot(x,U,'o'); hold    % Plot the computed solution

u=zeros(n-1,1);
for i=1:n-1,
  u(i) = cos(pi*x(i));
end
plot(x,u)    %%% Plot the true solution at the grid points
                on the same plot.


%%%%%%% Plot the error


figure(2); plot(x,U-u)


norm(U-u,inf)           %%% Print out the maximum error.
```

It is easy to check that the exact solution of the boundary value problem is $\cos(\pi x)$. If we plot the computed solution as defined by the values at the grid points (use plot(x,u,'o'), and the exact solution represented by the solid line in Figure 8.2 (a), the difference at the grid points is not too evident. However, if we plot the difference of the computed solution and the exact solution, which we call the error, we see there is indeed a small difference of $O(10^{-3})$, cf, Figure 8.2 (b). Note that, even if we do not know the true solution, we still can conclude that the error is $O(h^2)$ by theoretical analysis.

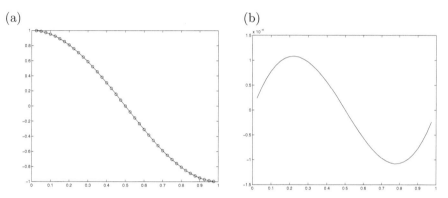

Figure 8.2. *(a) The plot of the computed solution (little 'o's), and the exact solution (solid line). (b) The plot of the error.*

Note that in Matlab, if use $[D, V] = eig(A)$, we get approximate eigenvalues in the diagonal matrix D, and corresponding eigenvectors in V.

8.2 Finite difference methods for 1D wave equations of BVPs

Since one-dimensional advection equations are easy to solve, we explain the finite difference method for one-dimensional wave equations with a source term,

$$\frac{\partial^2 u}{\partial t^2} = c^2 \frac{\partial^2 u}{\partial x^2} + f(x,t), \qquad 0 < x < L, \quad 0 < t < T,$$
$$u(0,t) = g_1(t), \qquad u(L,t) = g_2(t), \tag{8.9}$$
$$u(x,0) = f(x), \qquad \frac{\partial u}{\partial t}(x,0) = g(x), \qquad 0 < x < L.$$

It is not so easy to get analytic or series solution to the problem but it is quite simple using a finite difference method. As in the previous section, we set up a space grid first by selecting a parameter n, say $n = 100$, to get a space step size $h = \frac{L}{n}$, and a mesh,

$$x_i = ih, \quad i = 0, 1, \cdots, n.$$

Thus, we have $x_0 = 0$, $x_n = L$. Next we select a time step size Δt and get a grid in time,

$$t_k = k\Delta t, \quad k = 0, 1, \cdots, m, \quad \text{such that } n\Delta t = T.$$

Then we have a finite difference scheme to solve the initial-boundary value problem

$$U_i^0 = f(x_i), \quad U_i^1 = U_i^0 + \Delta t\, g(x_i), \quad i = 1, 2, \cdots, n-1, \quad \text{initial setup;}$$
$$U_0^{k+1} = g_1(t^{k+1}), \quad U_n^{k+1} = g_2(t^{k+1}), \quad \text{boundary conditions;}$$
$$\frac{U_i^{k-1} - 2U_i^k + U_i^{k+1}}{(\Delta t)^2} = c^2 \frac{U_{i-1}^k - 2U_i^k + U_{i+1}^k}{h^2} + f(x_i, t_k),$$
$$i = 1, 2, \cdots, n-1; \quad k = 1, 2, \cdots, m.$$
(8.10)

We can choose an m such that $\Delta t \leq \dfrac{h}{|c|}$ to ensure the stability of the finite difference method. Then we obtain a finite difference approximation to the solution at the final time T as

$$U_i^m \approx u(x_i, T), \quad i = 1, 2, \cdots n-1. \tag{8.11}$$

8.3 Finite difference methods for 1D heat equations

Similarly, we can use a finite difference method to solve a one-dimensional heat equation with a source term,

$$\begin{aligned}\frac{\partial u}{\partial t} &= c^2 \frac{\partial^2 u}{\partial x^2} + f(x, t), \quad 0 < x < L, \quad 0 < t < T, \\ u(0, t) &= g_1(t), \quad u(L, t) = g_2(t), \\ u(x, 0) &= f(x), \quad 0 < x < L.\end{aligned} \tag{8.12}$$

It is not so easy to get analytic or series solution to the problem but it is quite simple using a finite difference method. As in the previous section, we set up a space grid first by selecting a parameter n, say $n = 100$, to get a space step size $h = \frac{L}{n}$, and a mesh,

$$x_i = ih, \quad i = 0, 1, \cdots, n.$$

Thus, we have $x_0 = 0$, $x_n = L$. Next we select a time step size Δt and get a grid in time,

$$t_k = k\Delta t, \quad k = 0, 1, \cdots, m, \quad \text{such that } n\Delta t = T.$$

Then we have a finite difference scheme to solve the initial-boundary value problem

$$\begin{aligned}
U_i^0 &= f(x_i), \quad i = 1, 2, \cdots, n-1, \quad \text{initial setup;} \\
U_0^{k+1} &= g_1(t^{k+1}), \quad U_n^{k+1} = g_2(t^{k+1}), \quad \text{boundary conditions;} \\
\frac{U_i^{k+1} - U_i^k}{\Delta t} &= c^2 \frac{U_{i-1}^k - 2U_i^k + U_{i+1}^k}{h^2} + f(x_i, t_k), \\
i &= 1, 2, \cdots, n-1; \quad k = 0, 1, 2, \cdots, m.
\end{aligned} \quad (8.13)$$

We can choose an m such that $\Delta t \leq \dfrac{h^2}{2c^2}$ to ensure the stability of the finite difference method. Note that, in this case, the time step restriction is severe. There are variety of better methods around, see for example [7, 8, 14]. In the end, we obtain a finite difference approximation to the solution at the final time T as

$$U_i^m \approx u(x_i, T), \quad i = 1, 2, \cdots n-1. \tag{8.14}$$

8.4 Finite difference methods for 2D Poisson equations

We can also use a finite difference method to solve 2D or 3D Poisson equations or elliptic PDEs rather easily if the PDEs are defined on regular domains such as rectangles in 2D and cubics in 3D. After a finite difference discretization, a boundary value problem will be changed to a system of algebraic equations. We use a Poisson equations on a rectangle to explain the approach.

Let us now consider the following Poisson equation with a Dirichlet boundary condition:

$$u_{xx} + u_{yy} = f(x, y), \quad (x, y) \in \Omega = (a, b) \times (c, d), \tag{8.15}$$

$$u(x, y)|_{\partial\Omega} = u_0(x, y). \tag{8.16}$$

If $f \in C(\Omega)$, then the solution $u(x, y) \in C^2(\Omega)$ exists and it is unique. An analytic solution is often unavailable. A finite difference method is explained below.

- Step 1: Generate a grid. For example, a uniform Cartesian grid can be generated as,

$$x_i = a + ih_x, \quad i = 0, 1, 2, \cdots, m, \quad h_x = \frac{b-a}{m}, \tag{8.17}$$

$$y_j = c + jh_y, \quad j = 0, 1, 2, \cdots, n, \quad h_y = \frac{d-c}{n}. \tag{8.18}$$

In seeking an approximate solution U_{ij} at the grid points (x_i, y_j) where $u(x, y)$ is unknown, there are $(m-1)(n-1)$ unknowns.

- Step 2: Represent the partial derivatives with FD formulas involving the function values at the grid points. For example, if we adopt the three-point central FD formula for second-order partial derivatives in the x- and y-directions respectively, then

$$\frac{u(x_{i-1}, y_j) - 2u(x_i, y_j) + u(x_{i+1}, y_j)}{(h_x)^2} + \frac{u(x_i, y_{j-1}) - 2u(x_i, y_j) + u(x_i, y_{j+1})}{(h_y)^2}$$

$$\approx f_{ij}, \quad i = 1, \cdots m-1, \quad j = 1, \cdots n-1, \tag{8.19}$$

where $f_{ij} = f(x_i, y_j)$.

We replace the exact solution values $u(x_i, y_j)$ at the grid points with the approximate solution values U_{ij} obtained from solving the linear system of algebraic equations, i.e.,

$$\frac{U_{i-1,j} + U_{i+1,j}}{(h_x)^2} + \frac{U_{i,j-1} + U_{i,j+1}}{(h_y)^2} - \left(\frac{2}{(h_x)^2} + \frac{2}{(h_y)^2}\right) U_{ij} = f_{ij},$$

$$i = 1, 2, \cdots, m-1, \quad j = 1, 2, \cdots, n-1. \tag{8.20}$$

The FD equation at the grid point (x_i, y_j) involves five grid points in a five-point stencil, (x_{i-1}, y_j), (x_{i+1}, y_j), (x_i, y_{j-1}), (x_i, y_{j+1}), and (x_i, y_j). The grid points in the FD stencil are sometimes labeled as east, north, west, south, and the center in the literature. The center (x_i, y_j) is called the master grid point, where the FD equation is used to approximate the partial differential equation.

- Solve the linear system of algebraic equations (8.20), to get the approximate values for the solution at all of the grid points.

- Error analysis, implementation, visualization *etc.*

8.4.1 The matrix-vector form of the FD equations

In solving the algebraic system of equations by a direct method such as the Gaussian elimination or some sparse matrix technique, knowledge of the matrix-vector structure is important, although less so for an iterative solver such as the Jacobi, Gauss-Seidel or SOR(ω) methods. In the matrix-vector form $A\mathbf{U} = \mathbf{F}$, the unknown $\mathbf{U}$ is a 1D array. From 2D Poisson equations the unknowns U_{ij} are a 2D array, but we can order it to get a 1D array. We may also need to re-order the FD equations, and it is a common practice to *use the same ordering for the equations as for the unknown array*. There are two commonly used orderings, namely, the *natural ordering*, a natural choice for sequential computing; and *red-black ordering*, considered to be a good choice for parallel computing.

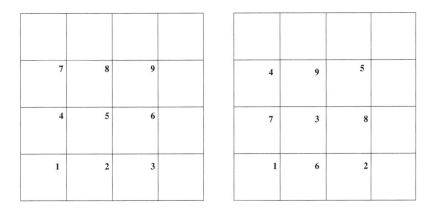

Figure 8.3. *The natural ordering (left) and the red-black ordering (right).*

The natural row ordering

In the natural row ordering, we order the unknowns and equations row by row. Thus the k-th FD equation corresponding to (i,j) has the following relation:

$$k = i + (m-1)(j-1), \quad i = 1, 2, \cdots, m-1, \quad j = 1, 2, \cdots, n-1, \quad (8.21)$$

see the left diagram in Figure 8.3.

Referring to Figure 8.3, suppose that $h_x = h_y = h$, $m = n = 4$. Then there are nine equations and nine unknowns, so the coefficient matrix is 9 by 9. To write down the matrix-vector form, use a 1D array $\mathbf{x}$ to express the unknown U_{ij}

according to the ordering, we should have

$$x_1 = U_{11}, \quad x_2 = U_{21}, \quad x_3 = U_{31}, \quad x_4 = U_{12}, \quad x_5 = U_{22},$$
$$x_6 = U_{32}, \quad x_7 = U_{13}, \quad x_8 = U_{23}, \quad x_9 = U_{33}.$$
(8.22)

Now if the algebraic equations are ordered in the same way as the unknowns, the nine equations from the standard central FD scheme using the five-point stencil are

$$Eqn.\ 1: \quad \frac{1}{h^2}(-4x_1 + x_2 + x_4) = f_{11} - \frac{u_{01} + u_{10}}{h^2}$$

$$Eqn.\ 2: \quad \frac{1}{h^2}(x_1 - 4x_2 + x_3 + x_5) = f_{21} - \frac{u_{20}}{h^2}$$

$$Eqn.\ 3: \quad \frac{1}{h^2}(x_2 - 4x_3 + x_6) = f_{31} - \frac{u_{30} + u_{41}}{h^2}$$

$$Eqn.\ 4: \quad \frac{1}{h^2}(x_1 - 4x_4 + x_5 + x_7) = f_{12} - \frac{u_{02}}{h^2}$$

$$Eqn.\ 5: \quad \frac{1}{h^2}(x_2 + x_4 - 4x_5 + x_6 + x_8) = f_{22}$$

$$Eqn.\ 6: \quad \frac{1}{h^2}(x_3 + x_5 - 4x_6 + x_9) = f_{32} - \frac{u_{42}}{h^2}$$

$$Eqn.\ 7: \quad \frac{1}{h^2}(x_4 - 4x_7 + x_8) = f_{13} - \frac{u_{03} + u_{14}}{h^2}$$

$$Eqn.\ 8: \quad \frac{1}{h^2}(x_5 + x_7 - 4x_8 + x_9) = f_{23} - \frac{u_{24}}{h^2}$$

$$Eqn.\ 9: \quad \frac{1}{h^2}(x_6 + x_8 - 4x_9) = f_{33} - \frac{u_{34} + u_{43}}{h^2}.$$

The corresponding coefficient matrix is *block tridiagonal*,

$$A = \frac{1}{h^2}\begin{bmatrix} B & I & 0 \\ I & B & I \\ 0 & I & B \end{bmatrix},$$
(8.23)

where I is the 3×3 identity matrix and

$$B = \begin{bmatrix} -4 & 1 & 0 \\ 1 & -4 & 1 \\ 0 & 1 & -4 \end{bmatrix}.$$

8.4. Finite difference methods for 2D Poisson equations

In general, for an $n+1$ by $n+1$ grid we obtain

$$A = \frac{1}{h^2}\begin{bmatrix} B & I & & \\ I & B & I & \\ & \ddots & \ddots & \ddots \\ & & I & B \end{bmatrix}_{n^2 \times n^2}, \quad B = \begin{bmatrix} -4 & 1 & & \\ 1 & -4 & 1 & \\ & \ddots & \ddots & \ddots \\ & & 1 & -4 \end{bmatrix}_{n^2 \times n^2}.$$

Since $-A$ is symmetric positive definite and weakly diagonally dominant, the coefficient matrix A is a non-singular matrix, and hence the solution of the system of the FD equations is unique.

The matrix-vector form is useful to understand the structure of the linear system of algebraic equations, and as mentioned it is required when a direct method (such as Gaussian elimination or a sparse matrix technique) is used to solve the system. However, it can sometimes be more convenient to use a two-parameter system, especially when an iterative method is preferred but also as more intuitive and to visualize the data.

8.4.2 The SOR(ω) iterative method

As we can see that the linear system of equations using a finite difference method to solve the Poisson equation is large ($O(n^2 \times n^2)$) but sparse with only $O(n^2)$ non-zero entries. One can use a simple iterative method, such as the Jacobi, Gauss-Seidel, or Successive Over Relaxation (SOR(ω)) method to solve the linear system of equations. Below is a description of the SOR(ω) iterative method given an initial guess $U_{ij}^{(0)}$ and $h_x = h_y = h$ so $M = N$. In the k-th iteration, $k = 0, 1, \cdots$, until the iteration converges, we set $U_{ij}^{(k+1)} = U_{ij}^{(k)}$, then

$$U_{ij}^{(k+1)} = (1-\omega)U_{ij}^{(k)} + \frac{\omega}{4}\left(U_{i-1,j}^{(k+1)} + U_{i+1,j}^{(k+1)} + U_{i,j-1}^{(k+1)} + U_{i,j+1}^{(k+1)} - h^2 f(x_i, y_j)\right)$$

$i, j = 1, 2, \cdots, n$.

(8.24)

The method works if $0 < \omega < 2$. When $\omega = 1$, the iterative method is called the Gauss-Seidel iterative method. The optimal *omega* for Poisson/Laplace equation is

$$\omega_{opt} = \frac{2}{1 + \sin(\pi h)} \sim \frac{2}{1 + \pi/n}. \tag{8.25}$$

A commonly use the stopping criterion is

$$\frac{\max_{ij}\left|U_{ij}^{(k+1)} - U_{ij}^{(k)}\right|}{\max_{ij}\left|U_{ij}^{(k+1)}\right|} < tol, \quad \text{say, } tol = 10^{-6}. \tag{8.26}$$

Brief summary

In this chapter, we presented a couple of examples using computers to solve differential equations. We can see that sometimes numerical methods do provide an effective alternative way in solving differential equations. We note that computational mathematics is a branch of mathematics that requires systematic study such as convergence, stability, accuracy, efficiency, reliability, etc., before we can take advantages of modern computer powers including artificial intelligence (AI).

8.5 Exercises

E8.1 Use a finite difference method to solve the advection equation:

$$\frac{\partial u}{\partial t} + a(x,t)u = f(x,t), \quad 0 < x < L, \quad a(x,t) \geq 0,$$

$$u(x,0) = u_0(x), \quad u(0,t) = g(t).$$

Try your method with $u(x,t) = \sin(x-2t)$ for $0 < t \leq 4$, $a(x) = x$, $0 < x < 2$, with $n = 32, 64, 128$. Other parameters and conditions can be determined from the give $u(x,t)$.

E8.2 Use a finite difference method to solve the wave equation:

$$\frac{\partial^2 u}{\partial t^2} = c^2 \frac{\partial^2 u}{\partial t^2} + f(x,t), \quad 0 < x < L,$$

$$u(x,0) = u_0(x), \quad \frac{\partial u}{\partial t} = v(x), \quad u(0,t) = g_1(t), \quad u(L,t) = g_2(t).$$

Try your method with $u(x,t) = \sin(x-2t)$ for $0 < t \leq 4$, $c = 5$, $0 < x < 2$, with $n = 32, 64, 128$. Other parameters and conditions can be determined from the give $u(x,t)$.

E8.3 Use a finite difference method to solve the heat equation:

$$\frac{\partial u}{\partial t} = a(x,t)\frac{\partial^2 u}{\partial t^2} + f(x,t), \quad 0 < x < L, \quad a(x,t) \geq 0,$$

$$u(x,0) = u_0(x), \quad u(0,t) = g_1(t), \quad u(L,t) = g_2(t).$$

8.5. Exercises

E8.4 Solve the Laplace equation on a unit square assuming that the true solution is (a): $u(x,t) = e^x \sin y$, (b): $u(x,t) = \sin(k_1\pi x)\cos(k_1\pi y)$. Use the source term and Dirichlet boundary condition from the given $u(x,y)$. Try different k_1 and k_2.

Appendix A

ODE review and other useful information

A.1 First order ODEs

For the readers' convenience, we review solutions or methods for solving some ordinary differential equations. We start with first order linear and homogeneous ODEs,

$$\frac{dy}{dx} + p(x)y(x) = 0. \tag{A.1}$$

For a more general ODE $a(x)\frac{dy}{dx} + b(x)y(x) = 0$ we can divide by $a(x)$ assuming it is not zero to get $\frac{dy}{dx} + \frac{b(x)}{a(x)}y(x) = 0$. If $a(x) = 0$ at some places, the differential equation is singular since there is no derivative involved.

If we rewrite the ODEs as

$$\frac{dy}{y} = -p(x)dx, \tag{A.2}$$

and integrate on both sides, we get

$$\int \frac{dy}{y} = -\int p(x)dx + C, \implies \log|y(x)| = C - \int p(x)dx.$$

The solution can be written as

$$y(x) = Ce^{-\int p(x)dx}. \tag{A.3}$$

For a non-homogeneous ODE

$$\frac{dy}{dx} + p(x)y(x) = g(x), \tag{A.4}$$

we can multiply a function $\mu(x)$, called an integrating factor so that the ODE can become something like $\frac{d}{dx}(sth.) = f(x)$ and can be integrated easily. In other words, after multiply an integrating factor to get

$$\mu(x)\frac{dy}{dx} + \mu(x)p(x)y(x) = g(x)\mu(x). \tag{A.5}$$

we wish the left hand side becomes $\frac{d}{dx}(\mu(x)y(x)) = g(x)\mu(x)$. Then the solution would be

$$y(x) = \frac{1}{\mu(x)}\left(\int g(x)\mu(x)dx + C_1\right).$$

The left hand side in (A.5) is the same as the left hand side of the ODE which leads to

$$\mu y' + \mu' y = \mu y' + \mu p y, \quad \text{or} \quad \mu' = \mu p.$$

We get $\mu(x) = C_2 e^{\int p(x)dx}$. Plugging this into (A.6), we get the solution

$$y(x) = \frac{1}{C_2}e^{-\int p(x)dx}\left(\int g(x)C_2 e^{\int p(x)dx} + C_1\right)$$
$$= e^{-\int p(x)dx}\left(\frac{C_1}{C_2} + \int g(x)e^{\int p(x)dx}\right)$$
$$= e^{-\int p(x)dx}\left(C + \int g(x)e^{\int p(x)dx}\right). \tag{A.6}$$

Example A.1. *Solve $y'(x) - y(x) = 2$.*

In this example, $p(x) = -1$, $g(x) = 2$, the solution is

$$y(x) = e^{\int 1 dx}\left(C + \int 2e^{\int (-1)dx}\right) = e^x\left(C - 2e^{-x}\right).$$

Note that, for this problem we can also use an undetermined coefficients method to find the particular solution $y_p = -2$. The solution then is the sum of the particular solution and the general solution to the homogeneous problem $y'(x) - y(x) = 0$.

A.2 Second order linear and homogeneous ODEs with constant coefficients

The ordinary differential equation has the form,

$$ay''(x) + b(x)y' + cy = 0. \tag{A.7}$$

The corresponding characteristic polynomial is defined as

$$a\lambda^2 + b\lambda + c = 0 \tag{A.8}$$

whose roots can be real or complex numbers depending on the discriminant,

$$b^2 - 4ac \begin{cases} > 0 & \text{there are two distinct roots } \lambda_1, \lambda_2; \\ = 0 & \text{there is one double root, } \lambda_1 = \lambda_2 = \lambda; \\ < 0 & \text{there are two complex roots } \lambda_1 = \alpha + \beta i, \lambda_2 = \alpha - \beta i, i = \sqrt{-1}. \end{cases}$$

The solution to the ODE are

- $b^2 - 4ac > 0$, $y(x) = C_1 e^{\lambda_1 x} + C_2 e^{\lambda_2 x}$.
- $b^2 - 4ac = 0$, $y(x) = C_1 e^{\lambda x} + C_2 x e^{\lambda x}$.
- $b^2 - 4ac < 0$, $y(x) = e^{\alpha x} (C_1 \cos \beta x + C_2 \sin \beta x)$.

A.3   Useful trigonometric formulas

There formulas below are useful in Fourier analysis, orthogonal expansions using trigonometric functions, and series solutions to partial differential equations.

$$\sin \alpha \cos \beta = \frac{1}{2} (\sin(\alpha + \beta) + \sin(\alpha - \beta)) \tag{A.9}$$

$$\cos \alpha \sin \beta = \frac{1}{2} (\sin(\alpha + \beta) - \sin(\alpha - \beta)) \tag{A.10}$$

$$\cos \alpha \cos \beta = \frac{1}{2} (\cos(\alpha + \beta) + \cos(\alpha - \beta)) \tag{A.11}$$

$$\sin \alpha \sin \beta = -\frac{1}{2} (\cos(\alpha + \beta) - \cos(\alpha - \beta)) \tag{A.12}$$

$$\sin^2 \alpha = \frac{1 - \cos 2\alpha}{2}; \quad \cos^2 \alpha = \frac{1 + \cos 2\alpha}{2} \tag{A.13}$$

$$\sin 2\alpha = 2 \sin \alpha \cos \alpha; \quad \cos 2\alpha = \cos^2 \alpha - \sin^2 \alpha. \tag{A.14}$$

A.4   ODE solutions to the Euler's equations

A second order Euler's ordinary differential equation has the following form

$$x^2 y'' + \alpha x y' + \beta y = 0. \tag{A.15}$$

An nth order Euler's ordinary differential equation has the following form

$$x^n \frac{d^n y}{dx^n} + a_{n-1} x^{n-1} \frac{d^{n-1} y}{dx^{n-1}} + \cdots a_1 xy + a_0 y = 0. \tag{A.16}$$

For a first order Euler equation $xy' + \alpha y = 0$, we have $\frac{y'}{y} = -\frac{\alpha}{x}$ and the solution is $\log |y(x)| = -\alpha \log |x| + C$ or $y = C|x|^\alpha$.

For a second order Euler's equation, we look for the solution of the form of $y(x) = x^r$, $y'(x) = rx^{r-1}$ and $y'(x) = r(r-1)x^{r-2}$. The ODE then becomes

$$x^2 r(r-1)x^{r-2} + \alpha x r x^{r-1} + \beta x^r = 0, \tag{A.17}$$

which leads to an indicial equation

$$r(r-1) + \alpha r + \beta = 0. \tag{A.18}$$

There are three cases corresponding to different general solutions.

1 Two distinct roots, r_1 and r_2. The solution then is

$$y(x) = C_1 |x|^{r_1} + C_2 |x|^{r_2}.$$

2 One repeated root r_1. The solution is

$$y(x) = C_1 |x|^{r_1} + C_2 \left(\log |x| \right) |x|^{r_1}.$$

3 A complex pair $r = a \pm ib$, the solution is

$$y(x) = |x|^a \left(C_1 \cos(b \log |x|) + C_2 \sin(b \log |x|) \right).$$

Example A.2. *Solve the ordinary differential equation* $x^2 y'' + 2xy' - 6y = 0$.

Solution: The indicial equation is $r(r-1) + 2r - 6 = 0$, or $r^2 + r - 6 = (r+3)(r-2) = 0$. Its roots are $r_1 = -3$ and $r_2 = 2$. The general solution is

$$y(x) = C_1 |x|^{-3} + C_2 |x|^2 = \frac{C_1}{|x|^3} + C_2 x^2.$$

Example A.3. *Solve the ordinary differential equation* $x^2 y'' + 3xy' + 10y = 0$.

Solution: The indicial equation is $r(r-1) + 3r + 10 = 0$, or $r^2 + 2r + 10 = 0$. The solutions are $r = -1 \pm 3i$. The general solution is

$$y(x) = |x|^{-1} \left(C_1 \cos(3 \log |x|) + C_2 \sin(3 \log |x|) \right).$$

A.4. ODE solutions to the Euler's equations

Example A.4. *Solve the ordinary differential equation $x^2 y'' + 3xy' + y = 0$.*

The indicial equation is $r(r-1) + 3r + 1 = 0$, or $r^2 + 2r + 1 = 0$. There is one double root $r = -1$. The general solution is

$$y(x) = C_1 |x|^{-1} + C_2 \left(\log |x|\right) |x|^{-1}.$$

Appendix B

Introduction to Maple

Maple is a computer software package produced by Maplesoft, Inc in Waterloo, Ontario, Canada. It is a programming language as well as a powerful computer algebra system that is well-suited for use in a course on partial differential equations. Here we provide the student with an introduction that should suffice for the various uses (computations, graphing and animations) that are needed throughout this textbook.

B.1 The Maple worksheet

One "Works with Maple" in what is referred to as a Maple worksheet. Below is a screen shot of a Maple worksheet that illustrates the two types of "Lines" in a Maple worksheet: (1) command lines, and (2) text lines, see Figure B.1.

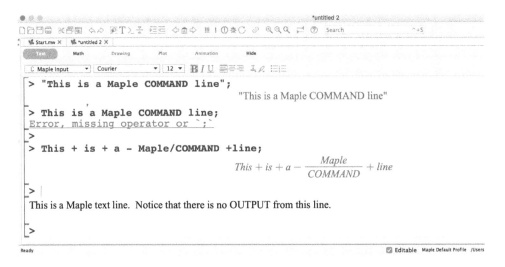

Figure B.1. *Maple command and text lines.*

In the worksheet pictured above the first line with the words "This is a Maple COMMAND line"; is in fact a "command" line. Command lines are also known as Execution groups in Maple. Note that command lines are indicated by the > symbol. We insert Maple commands into a command line to compute, or to graph, or perhaps to create an animation. We will provide examples of all of these operations below. Since the words in the first line are enclosed in quotation marks, Maple considers "This is a Maple COMMAND line" to be a maple string, and when "enter" is pressed, spits out the words of the string in blue as shown.

The second line has the same words as the first line, but without quotation marks. Upon execution Maple replies with the warning:

$$\text{Error, missing operator or ';'}$$

This is because maple is looking for operators, such as $*$, $/$, $+$ between the words. In the third line we've inserted some operators between the words just to illustrate that Maple now considers the individual words to be valid Maple objects and dutifully tries to perform the (meaningless) operations. Please also note that each line in a

B.1. The Maple worksheet

Maple worksheet should end with a semi-colon (output shown) or a colon (command executed by output suppressed).

Finally, the last line with text in it is a text line. There is no $>$ symbol indicating a textline. No commands can be executed inside a textline. Textlines are only for text.

SUMMARY:. Maple command lines are indicated with the symbol $>$, and one enters Maple commands in command lines. Pressing "enter" after inserting a command in a command line will result in Maple executing the command. Maple text lines are not indicated using the symbol $>$ - rather they are simply white spaces ready for text input.

Adding extra command lines

If can happen that you are working at some point in a Maple worksheet, and you realize that you need to have another command line at that point in your work. To insert a new Execution Group at a point in the worksheet, put the cursor at the spot where you want to make the insertion. Then with your mouse click on "insert" in the top menu, then click on Execution Group in that menu, and then choose between "before cursor" and "after cursor". See Figure B.2.

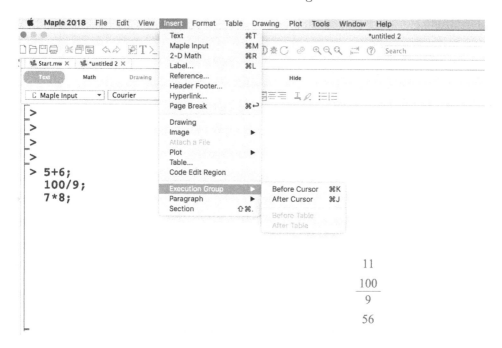

Figure B.2. *Adding a new command line.*

Multiple commands in a single execution group

If can often happen that you want to define and then execute several commands in sequence. You can, of course, simply put your sequence of commands in separate command lines and then execute them one at a time. On the other hand, you can input mutliple commands in a single execution group. You create room in an execution group by entering shift-return from the key board. This will insert 1 extra line. More lines in the same execution group will be produced each time you press shift-return on the key board. In Figure B.2 above the last execution group has 3 Maple commands in one execution group.

Adding extra text lines

If can happen that you are working at some point in a Maple worksheet, and you realize that you need to have another text line at that point in your work. To insert a new text line at a point in the worksheet, put the cursor in a command line at the spot where you want to make the insertion. Then with your mouse click on "insert" in the top menu and then click on Text in that menu. Maple will convert the command line to a text line.

B.2 Arithmetic and algebraic operations in Maple

Maple is a computational workhorse that is capable of computing almost anything a PDE-student needs. The 4 basic operations of addition, subtraction, multiplication and division are denoted by commands similar to the commands on many hand-held calculators. They are:

Addition Place the + sign between two things you want to add together. For example, $\boxed{5+6}$.

Subtraction To subtract A from B use $\boxed{B-A}$.

Multiplication To multiple A and B use $\boxed{A*B}$

Division To divide A by B use $\boxed{A/B}$

It is important to use proper syntax when using operations in combinations. Maple uses parentheses ( ), and only parentheses, to organize these operations when used together. That is to say one may not use curly braces { }, or square braces [ ] to organize these operations. This is because Maple uses { } to define sets, and [ ] to define lists. The difference between a Maple set and a Maple list is that the elements of a set are unique, while elements of a list need not be unique. Moreover, lists are ordered so that the order you define elements of a list is preserved. Maple will often reorder the elements of a set using an ordering determined by Maple. Set Figure B.3 to see how these work.

B.2. Arithmetic and algebraic operations in Maple

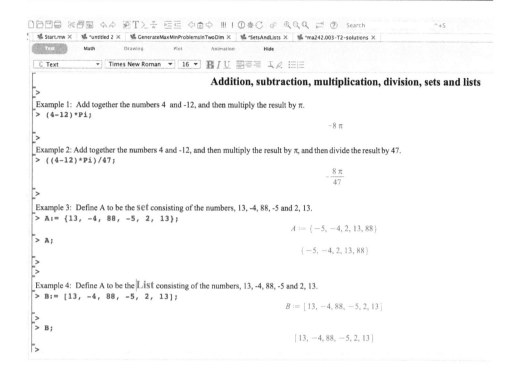

Figure B.3. *Operations, sets and lists.*

Notice in the examples that to enter π in a Maple worksheet you must use Pi and not pi. Also notice that in defining A to be a set, the number 13 is entered twice but only shows up once, since elements of a set are unique. On the other hand, in defining B to be a list, the order is preserved and 13 appears twice.

B.2.1 The assignment operator

Also illustrated in Figure B.3 is the use of the assignment operator. The assignment operator is denoted by the combination of a colon followed by an equal sign, namely := preceded by the name you want to assign to what is on the right-hand-side. The letter A is assigned to the set with elements 13, -4, 88, -5 and 2, 13 while the letter B is assigned to the list with ordered elements 13, -4, 88, -5 and 2, 13. Once you assign a value to a name, Maple remembers that assignment until you close the worksheet or issue a "restart" command. This is illustrated in Examples 3 and 4 in Figure B.3 above.

B.2.2 Special symbols in Maple

Below is a short list of the special symbols used in a Maple worksheet.

1. π. The symbol π, namely the ratio of the circumference of a circle to its diameter, is entered in a Maple worksheet as Pi.

2. e. The real number that is denoted as e, is by definition that real number for which the following limit holds

$$\lim_{h \to 0} \frac{e^h - 1}{h} = 1$$

   It evaluates to 10 digits to the number 2.718281828 as shown in Figure B.4. Note the first line in the figure, showing that e is just e in a Maple worksheet. To evaluate it we must use exp(1).

```
> evalf(e);
                                              e
> evalf(exp(1));
                                      2.718281828
>
```

Figure B.4. *Evaluate the exponential function.*

3. The complex number $\sqrt{-1}$ is denoted by the capitol letter I in the Maple worksheet.

B.2.3 Getting started with Maple

It is time to try out these ideas in a Maple worksheet. First download the file A-maple-worksheet and save it to disk. It is a Maple worksheet verison of this pdf. Next start up Maple and use the file command to open the worksheet, as shown in the figure below. You'll need to find the file you saved on your disk. When the file opens you can follow along with the Maple worksheet and this pdf, see Figure B.5.

Figure B.5. *Opening a Maple worksheet.*

B.3 Functions in Maple

In working with Fourier series and eigenfunction expansions it is useful to be able to define certain functions to help with the calculations. The syntax for defining a function and assigning a name to the function in Maple is illustrated in Figure B.6.

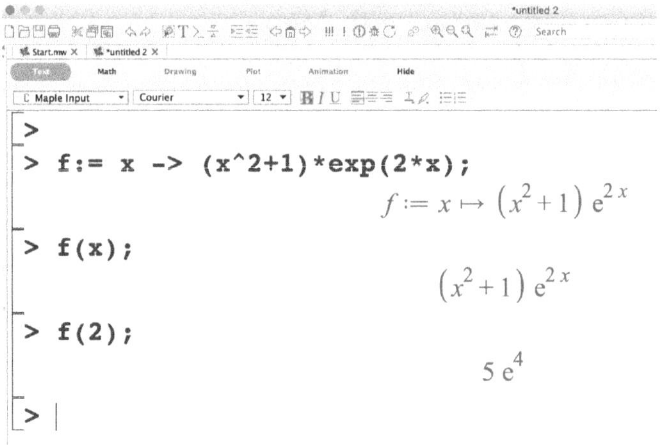

```
> 
> f:= x -> (x^2+1)*exp(2*x);
```
$$f := x \mapsto \left(x^2+1\right) e^{2x}$$
```
> f(x);
```
$$\left(x^2+1\right) e^{2x}$$
```
> f(2);
```
$$5 e^4$$
```
> |
```

Figure B.6. *Functions in Maple.*

In the first line in the figure we are "assigning f to be the function that sends x to the value $(x^2+1)\exp(2*x)$". The "send to" is denoted by –>, which is a "dash" followed by the "greater than" symbol >. Once the assignment is made, we call the function using standard function notation. In the second line in Figure B.6 we write $f(x)$ to "evaluate f at x" to produce the value of the function at an arbitrary x. In the third line we we write $f(2)$ to "evaluate f at the number 2" to get the value shown, namely $5e^4$.

B.3.1 Built in functions in Maple

Maple provides a number of predefined functions to work with in the Maple worksheet. A short list of the useful predefined functions are

1 exp. The exponential function is called in the Maple worksheet as shown in Figure B.7.

B.3. Functions in Maple

```
> exp(x);
  exp(1);
  exp(x^2-2*x+1);
```

$$e^x$$

$$e$$

$$e^{x^2-2x+1}$$

```
> |
```

Figure B.7. *Built in functions in Maple.*

REMARK:. It is important to note that one must use the exponential function using the exp() format. Using the notation e^x in a Maple worksheet denotes the letter e with exponent x, and does not represent the exponential function.

2 ln. The natural logarithm function is called in the Maple worksheet as shown in Figure B.8.

```
> ln(x);
  ln(1);
  ln(exp(1));
  ln(x^2-2*x+1);
```

$$\ln(x)$$

$$0$$

$$1$$

$$\ln(x^2 - 2x + 1)$$

```
> |
```

Figure B.8. *Natural logarithm function in Maple.*

3 The trig and inverse trig functions sin, cos,tan,cot,sec,csc, and their inverses arcsin, arccos,arctan,arccot,arcsec,arccsc are predefined in Maple.

4 The hyperbolic trig and inverse hyperbolic trig functions sinh, cosh,tanh,coth, sech,csch, and their inverses arcsinh, arccosh,arctanh,arccoth,arcsech, arccsch are predefined in Maple.

5 evalf(). Maple's command that evaluates valid expressions to N digits using floating point arithmetic is the evalf() command. Here N is a positive in-

teger and valid expressions are numbers, for example rational numbers, valid symbols, for example π, and Maple functions such as those listed above. The syntax to evaluate a valid expression to N decimal places is

$\Big[$ >evalf("valid expression", N)

As an example, Pi evaluates to 100 digits to the number, see Figure B.9.

```
> evalf(Pi,100);
    3.1415926535897932384626433832795028841971693993751058209749445923078164062862089986280348253421170 68
>
>
```

Figure B.9. *Evaluating π in Maple.*

B.3.2 Plotting functions in a Maple worksheet

Consider the function $f := x \to \sin(3\pi x)$. The domain is the entire real line, and to display the graph of f over an interval, say $x = a \ldots b$, we use the plot command. The syntax for our plot is

$\Big[$ > plot $\Big(f(x), x = a..b, \text{options}\Big)$

where "options" refers to the plot options available in Maple, including color and thickness. The color option is quite useful. For example, to color a curve blue we use the command

$\Big[$ > plot $\Big(f(x), x = a..b, \text{color=blue}\Big)$

To plot two functions $f(x)$ and $g(x)$ on the same plot, we group $f(x)$ and $g(x)$ together into the list $[f(x), g(x)]$ and use this list as the first argument in the plot command.

$\Big[$ > plot $\Big([f(x),g(x)], x = a..b, \text{color=[blue,red]}\Big)$

REMARK:. Because lists are ordered, Maple knows that in executing the above command the curve $y = f(x)$, the first element in the input list $[f(x), g(x)]$, should should be plotted in blue, the first element in the color list $.color = [blue, red]$.

B.3. Functions in Maple

Similarly, the curve $y = g(x)$ should be plotted in red. See Figure B.10 where the option on "thickness=n" of the curves is added, where n is a non-negative integer.

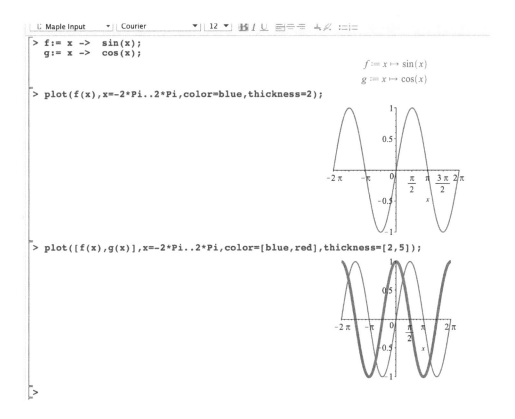

Figure B.10. *Plotting in Maple.*

B.3.3 Animating plots

Maple has the ability to produce animations of plots. For example, consider the function f defined by

$$> f := (x,t) -> \sin(x - t)$$

for $0 \leq x \leq 2\pi$ and $0 \leq t \leq 10$. Looking at the plot of f in the xy-plane, we see a single sine wave moving from left to right across the screen as t advances. A few snap shots are shown in Figure B.11. However in the maple worksheet the animation is live. Download the Maple worksheet containing this example HERE.

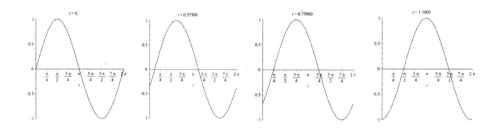

Figure B.11. *Snap plots in Maple.*

The command to produce the animation is contained in the **plots** package and is the animate() command. It is called with the format

$$> \text{with(plots);}$$

$$> \text{animate(plot,[sin(x-y),x=0..2*Pi], t=0..10);}$$

When these commands are executed Maple responds with a picture of the plot as shown in Figure B.12

B.3. Functions in Maple

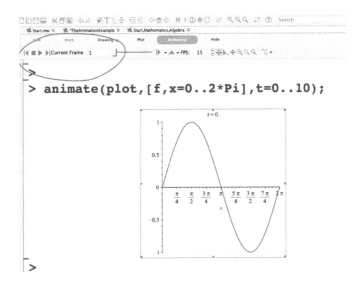

Figure B.12. *Animating plots in Maple.*

Clicking on the plot with the cursor will cause Maple to create the "movie player controls" shown inside the red circle in Figure B.12. Click on the play button ▶ to play the animation.

B.3.4 Differentiating functions in a Maple worksheet

The syntax for differentiating a function g that has been defined in the worksheet is

$$> \mathbf{diff(g(x),x)}$$

See Figure B.13. g is defined in the first line, and the second line is there to emphasize that you must use $g(x)$ and not g when working with the function. The derivative of g is given in the third line in the figure.

A second derivative uses the notation

$$> \mathbf{diff(g(x),x,x)}$$

Differentiating multivariable functions

There is a similar notation for partial derivatives of functions of more than one variable. Suppose that f is defined in the worksheet as a function of x and y as in the fourth input line in Figure B.13. In the execution group beginning with the fifth input line, the following partial derivatives are computed.

$$\frac{\partial f}{\partial x},\quad \frac{\partial^2 f}{\partial y \partial x},\quad \frac{\partial f}{\partial y},$$

The second and third input lines in Figure B.13 illustrate an important point. Once g is assigned as a function, you cannot use g as the argument of the diff(), or any other Maple command; you must use $g(x)$ for the values of g at x.

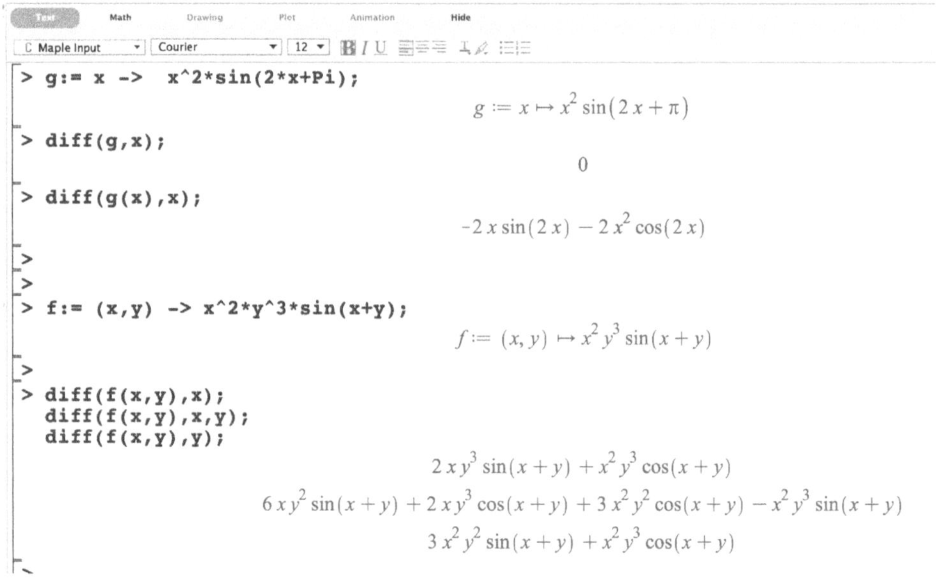

```
> g:= x -> x^2*sin(2*x+Pi);
```
$$g := x \mapsto x^2 \sin(2x + \pi)$$

```
> diff(g,x);
```
$$0$$

```
> diff(g(x),x);
```
$$-2x\sin(2x) - 2x^2\cos(2x)$$

```
>
>
> f:= (x,y) -> x^2*y^3*sin(x+y);
```
$$f := (x, y) \mapsto x^2 y^3 \sin(x+y)$$

```
>
> diff(f(x,y),x);
  diff(f(x,y),x,y);
  diff(f(x,y),y);
```
$$2xy^3 \sin(x+y) + x^2 y^3 \cos(x+y)$$
$$6xy^2 \sin(x+y) + 2xy^3 \cos(x+y) + 3x^2 y^2 \cos(x+y) - x^2 y^3 \sin(x+y)$$
$$3x^2 y^2 \sin(x+y) + x^2 y^3 \cos(x+y)$$

Figure B.13. *Differentiation in Maple.*

B.3.5 Integrating functions in a Maple worksheet

Suppose f is a continuous function of one variable defined in a Maple worksheet. See, for example, the definition of $f(x) = x\cos(x)$ in Figure B.14. The syntax for integrating f over an interval $x = a..b$ using the int() command is

$$> \text{int}\Big(\text{f(x)},\text{x=a..b}\Big)$$

Choosing a specific interval, say $x = 2..10$ we can integrate f over this interval as shown in the, second input line in Figure B.14. Note that in the third line in the figure that "int" has been changed to "Int", which is referred to as the "inert version of int" in that the command is simply printed out and not executed. This is useful for the student to make certain that "what was typed in is actually what you wanted typed in and not a mistake."

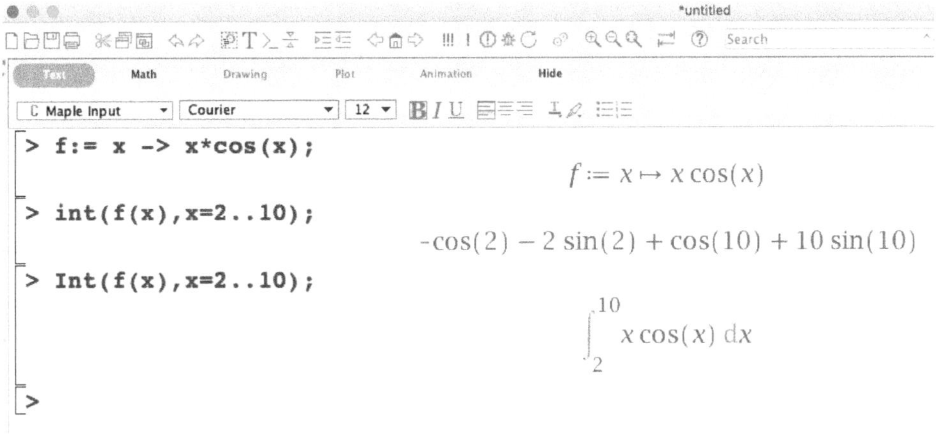

```
> f:= x -> x*cos(x);
                    f := x ↦ x cos(x)
> int(f(x),x=2..10);
                    -cos(2) - 2 sin(2) + cos(10) + 10 sin(10)
> Int(f(x),x=2..10);
                    ∫₂¹⁰ x cos(x) dx
>
```

Figure B.14. *Integration in Maple.*

B.3.6 Functions defined on integers

A standard computation in Fourier series and eigenfunction expansions is to define a function which is to be evaluated on integers. Maple again uses the notation

> f:=x → expression in x

but simply replaces the variable x with a variable n which will be used as an integer. An example is shown in Figure B.15 where the name a is assigned to the function that sends an integer n to the value of the integral of the function $f(x)\cos(n\pi x/L)$ from $x = 0$ to $x = L = 1$. The values $a(n)$ for $n = 1, 4, 10$ are then printed out.

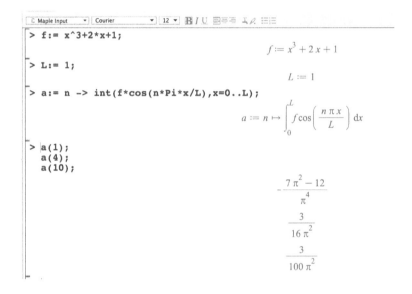

Figure B.15. *Function defined on integers in Maple.*

B.3.7 Partial sums

It is useful in dealing with infinite series to create partial sum approximations of the full series. Thus for the infinite series

$$S := \sum_{i=1}^{\infty} f(i)$$

the Nth-partial sum, $N \geq 1$, is the finite sum S_N given by

$$S_N = \sum_{i=1}^{N} f(i)$$

B.3. Functions in Maple

For example, a **Fourier sine series** of of a function $f(x)$ on the interval $[0, L]$ has the form

$$FS = \sum_{n=1}^{\infty} b(n) \sin(n\pi x/L) \tag{B.1}$$

where the coefficients are defined by

$$b := n \to \frac{2}{L} \int_0^L f(x) \sin(n\pi x/L) dx. \tag{B.2}$$

The Nth-partial sum, $N \geq 1$, for such a Fourier sine series is simply

$$S_N = \sum_{n=1}^{N} b(n) \sin(n\pi x/L)$$

In the Maple worksheet we use the sum() command to create partial sums of functions defined on integers. The syntax is, for a function f defined on integers,

```
> sum(f(i), i=1..N) ;
```

where N is a positive integer. In Figure B.16 the partial sums for a Fourier sine series of a function is illustrated, followed by a plot of the partial sums S_N for $N = 1, 2, 3, 20$ in Figure B.17.

```
> f:= x ->x^3+2*x+1;
```
$$f := x \mapsto x^3 + 2x + 1$$

```
> L:= 1;
```
$$L := 1$$

```
> b:= n -> 2/L*Int(f(x)*sin(n*Pi*x/L),x=0..L);
  "The Fourier Sine Series"= Sum(b(n)*sin(n*Pi*x/L),n=1..infinity);
  S:= N -> sum(b(n)*sin(n*Pi*x/L),n=1..N);
```

$$b := n \mapsto \frac{2\left(\int_0^L f(x) \sin\left(\frac{n\pi x}{L}\right) dx\right)}{L}$$

$$\text{"The Fourier Sine Series"} = \sum_{n=1}^{\infty} 2\left(\int_0^1 f(x) \sin(n\pi x)\, dx\right) \sin(n\pi x)$$

$$S := N \mapsto \sum_{n=1}^{N} b(n) \sin\left(\frac{n\pi x}{L}\right)$$

```
> S(1);
```
$$\frac{2\left(5\pi^3 - 6\pi\right) \sin(\pi x)}{\pi^4}$$

```
> S(2);
```
$$\frac{2\left(5\pi^3 - 6\pi\right) \sin(\pi x)}{\pi^4} + \frac{\left(-24\pi^3 + 12\pi\right) \sin(2\pi x)}{8\pi^4}$$

```
> S(3);
```
$$\frac{2\left(5\pi^3 - 6\pi\right) \sin(\pi x)}{\pi^4} + \frac{\left(-24\pi^3 + 12\pi\right) \sin(2\pi x)}{8\pi^4} + \frac{2\left(135\pi^3 - 18\pi\right) \sin(3\pi x)}{81\pi^4}$$

Figure B.16. *Partial sums of a series in Maple.*

```
> plot([f(x),S(1),S(2),S(3),S(20)],x=0..1,color=[black,red,blue,green,brown]);
```

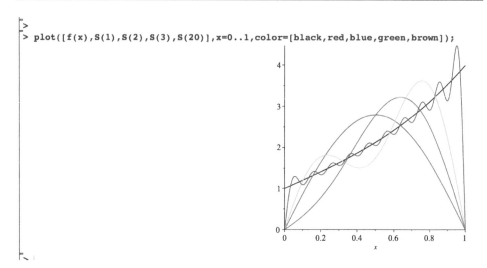

Figure B.17. *Plot of partial sums S_N for $n = 1,2,3,20$.*

B.3.8 Maple's repetition statement

To quote from Maple's help page, *The repetition statement provides the ability to execute a statement sequence repeatedly, either for a counted number of times (using the for...to clauses) or until a condition is satisfied (using the while clause). Both forms of clauses can be present simultaneously.* It strings together the words "for ... by ... to ... while". Below is an example call to the repetition statement.

> **TheValue := 1;**

    for i from 1 by 1 while TheValue $< 1.0 * 10^{19}$

    do

        "i"=i;

        TheValue:= evalf(exp(i));

    end do;

In this example we are asked to find the first positive integer i such that $\exp(i) > 1.0 * 10^{19}$. In the first line the variable name TheValue is first initialized to the value 1. Then in the loop we evaluate the exponential function $\exp(i)$ at the integer i, where i starts at 1 and increments **by 1**. The loop will stop once TheValue surpasses the value $\mathbf{1.0 * 10^{19}}$ due to the phrase while TheValue $< 1.0 * 10^{19}$.

Notice that in Figure B.18 the output is shown for each step since the line end do; ends with a semi-colon. For large calculations in a loop you may want to suppress the output until the loop ends. This is illustrated in Figure B.19 where end do: ends with a colon. We then see the answer. The end of the loop prints out which integer (44) with the line The_i, and then with the line TheValue prints out the 10-digit approximation $1.285160011 \times 10^{19}$ of exp(44).

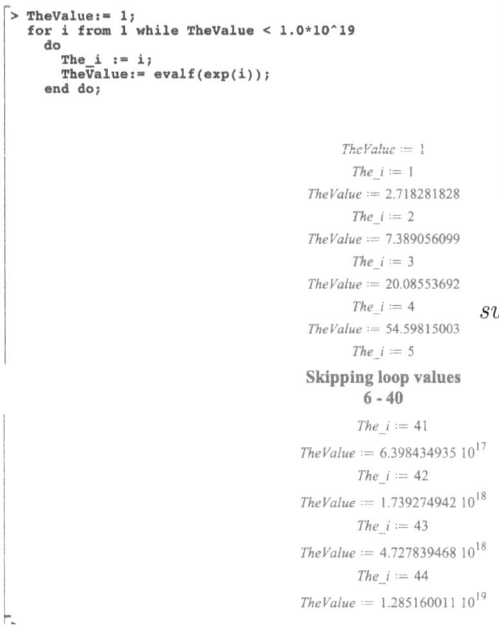

```
> TheValue:= 1;
  for i from 1 while TheValue < 1.0*10^19
  do
      The_i := i;
      TheValue:= evalf(exp(i));
  end do;
```

$TheValue := 1$
$The\_i := 1$
$TheValue := 2.718281828$
$The\_i := 2$
$TheValue := 7.389056099$
$The\_i := 3$
$TheValue := 20.08553692$
$The\_i := 4$
$TheValue := 54.59815003$
$The\_i := 5$

Skipping loop values 6 - 40

$The\_i := 41$
$TheValue := 6.398434935 \; 10^{17}$
$The\_i := 42$
$TheValue := 1.739274942 \; 10^{18}$
$The\_i := 43$
$TheValue := 4.727839468 \; 10^{18}$
$The\_i := 44$
$TheValue := 1.285160011 \; 10^{19}$

Figure B.18. *A loop in Maple.*

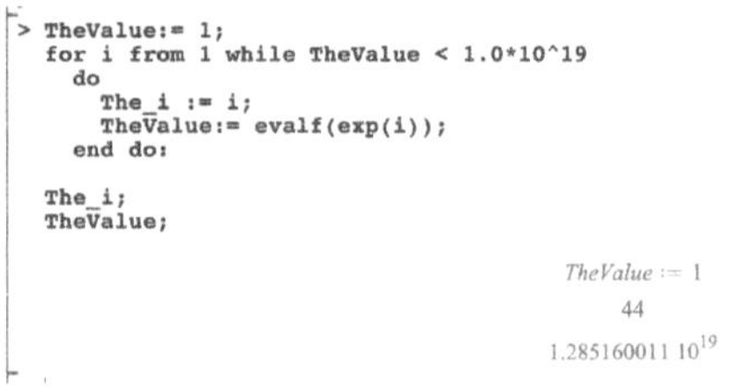

```
> TheValue:= 1;
  for i from 1 while TheValue < 1.0*10^19
  do
      The_i := i;
      TheValue:= evalf(exp(i));
  end do:
  The_i;
  TheValue;
```

$TheValue := 1$
44
$1.285160011 \; 10^{19}$

Figure B.19. *Loop with output suppressed.*

B.4 Computing sine, cosine, and full Fourier series

B.4.1 Fourier sine series

The Fourier sine series (SS) of a function $f(x)$ on the interval $[0, L]$ has the form

$$SS = \sum_{n=1}^{\infty} b(n) \sin(n\pi x/L) \tag{B.3}$$

where the coefficients are defined by

$$b := n \to \frac{2}{L} \int_0^L f(x) \sin(n\pi x/L) dx, n = 1, 2, \ldots \tag{B.4}$$

The Nth-partial sum, $N \geq 1$, for such a Fourier sine series is simply

$$SS\_Partial\_Sum = \sum_{n=1}^{N} b(n) \sin(n\pi x/L)$$

Thus to compute the Fourier SS of a given function $f(x)$ we need:

B.4. Computing sine, cosine, and full Fourier series

1 The function $f(x)$, and

2 The number L defining the domain $x = 0 \ldots L$ of the problem.

In Figure B.19 we show how to define the Fourier series partial sum function for the function e^x on the interval $[0,1]$. In the figure the definition of the SS coefficients function $b(n)$ as well as the definition of the SS_Partial_Sum function are shown. Then the plot is shown for e^x and the 20-term partial sum approximation of the sine series for e^x.

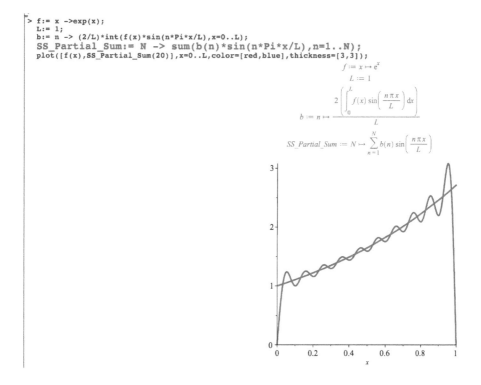

Figure B.19. *Define the Fourier series partial sum function.*

B.4.2 Fourier cosine series

The Fourier cosine series (CS) of a function $f(x)$ on the interval $[0, L]$ has the form

$$CS = \frac{a_0}{2} + \sum_{n=1}^{\infty} a(n) \cos(n\pi x/L) \tag{B.5}$$

where the coefficients are defined by

$$a := n \to \frac{2}{L} \int_0^L f(x) \cos(n\pi x/L) dx, n = 0, 1, 2, \ldots \tag{B.6}$$

The Nth-partial sum, $N \geq 1$, for such a Fourier cosine series is simply

$$CS\_Partial\_Sum = \frac{a_0}{2} + \sum_{n=1}^{N} a(n) \cos(n\pi x/L)$$

Thus to compute the CS of a given function $f(x)$ we need:

1 The function $f(x)$, and

2 The number L defining the domain $x = 0 \ldots L$ of the problem.

In Figure B.20 we show how to define the Fourier cosine series partial sum function for the function $x^3 - 2x^2 + x + 1$ on the interval [0,1]. In the figure the definition of the CS coefficients function $a(n)$ as well as the definition of the CS_Partial_Sum function are shown. Then the plot is shown for $x^3 - 2x^2 + x + 1$ and the 5-term partial sum approximation of the cosine series for $x^3 - 2x^2 + x + 1$.

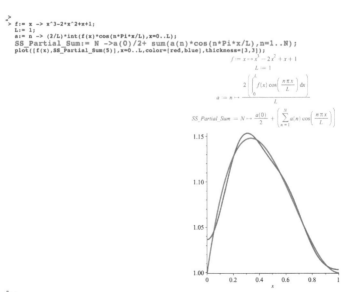

Figure B.20. *A cosine series example.*

B.4.3 Fourier series

The Fourier series (FS) of a function $f(x)$ on the interval $[-L, L]$ has the form

$$FS = \frac{a_0}{2} + \sum_{n=1}^{\infty} \Big(a(n) \cos(n\pi x/L) + b(n) \sin(n\pi x/L) \Big) \tag{B.7}$$

where the coefficients are defined by

$$a := n \to \frac{1}{L} \int_{-L}^{L} f(x) \cos(n\pi x/L) dx, \, n = 0, 1, 2, \ldots \tag{B.8}$$

$$b := n \to \frac{1}{L} \int_{-L}^{L} f(x) \sin(n\pi x/L) dx, \, n = 1, 2, \ldots \tag{B.9}$$

The Nth-partial sum, $N \geq 1$, for such a Fourier series is simply

$$FS\_Partial\_Sum = \frac{a_0}{2} + \sum_{n=1}^{N} \Big(a(n) \cos(n\pi x/L) + b(n) \sin(n\pi x/L) \Big).$$

Thus to compute the FS of a given function $f(x)$ we need:

1 The function $f(x)$, and

2 The number L defining the domain $x = -L \ldots L$ of the problem.

In Figure B.21 we show the definition of the FS coefficients functions $a(n)$ and $b(n)$, and then the construction of the Nth-partial sum function for this FS. The figure also contains a plot of the function $f(x) = x^3 - 2x + x + 1$ and the 10-term Partial sum approximation of the FS for $f(x)$.

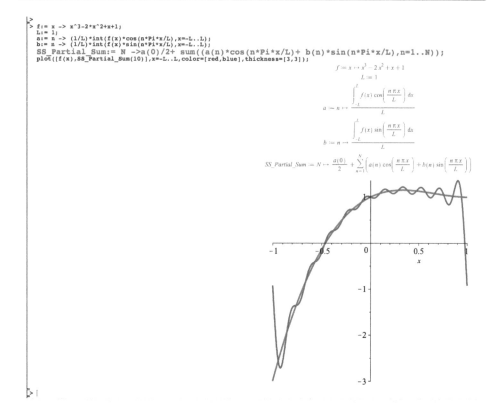

Figure B.21. *A full Fourier series expansion in Maple.*

B.5 Families of orthogonal functions in Maple

In your study of partial differential equations and boundary value problems you will encounter special functions defined by the PDE and the BCs. An important definition is the following.

Definition B.1. *An orthogonal family of functions on an interval $[a,b]$ with weight function $w(x)$ is a set of nontrivial functions $\{X_n(x)\}_{n=1}^{\infty}$ whose members satisfy the equations*

$$\int_a^b w(x) X_m(x) X_n(x) dx = 0 \quad \text{for } m \neq n.$$

B.5.1 The family of sine functions $\{\sin(n\pi x/L)\}_{n=1}^{\infty}$

The family of sine functions $\{\sin(n\pi x/L)\}_{n=1}^{\infty}$ satisfies

$$\int_0^L \sin(m\pi x/L) \sin(n\pi x/L) dx = 0 \quad \text{for } m \neq n$$

and hence is an orthogonal family on the interval $[0, L]$ with weight function $w(x) = 1$.

B.5.2 The family of functions $\{1, \cos(n\pi x/L)\}_{n=1}^{\infty}$

The family of functions $\{1, \cos(n\pi x/L)\}_{n=1}^{\infty}$ satisfies

$$\int_0^L \cos(m\pi x/L) \cos(n\pi x/L) dx = 0 \quad \text{for } m \neq n \ \ m,n = 0, 1, 2, \ldots$$

and hence is an orthogonal family on the interval $[0, L]$ with weight function $w(x) = 1$.

B.5.3 The family of functions $\{1, \cos(n\pi x/L), \sin(n\pi x/L)\}_{n=1}^{\infty}$

The family of functions $\{1, \cos(n\pi x/L), \sin(n\pi x/L)\}_{n=1}^{\infty}$ is an orthogonal family of functions on the interval $[-L, L]$ with weight function $w(x) = 1$ since they satisfy

$$\int_{-L}^L X_m(x) X_n(x) dx = 0 \quad \text{for } m \neq n$$

where $X_m(x)$ and $X_n(x)$ are distinct elements in the set of functions.

B.5.4 The Legendre polynomials

Let $P_n(x)$, $n = 0, 1, 2, \ldots$ denote the bounded solutions of Legendre's differential equation
$$(1 - x^2)y'' - 2xy' + n(n+1)y = 0$$
with $-1 \le x \le 1$. These solutions are even and odd polynomials. For example, $P_0(x) = 1$, $P_1(x) = x$ and $P_2(x) = \frac{1}{2}(3x^2 - 1)$. This family of Legendre polynomials is an orthogonal family on the interval $[-1, 1]$ with weight function $w(x) = 1$ since
$$\int_{-1}^{1} P_m(x) P_n(x) dx = 0 \quad \text{for} \quad m \ne n$$

On the other hand when $m = n$ we have
$$\int_{-1}^{1} \left(P_n(x)\right)^2 dx = \frac{2}{2n+1}$$

See the Maple worksheet Legendre Polynomials in Maple to learn how to use Legendre Polynomials in Maple.

B.5.5 The Bessel functions

As discussed in Section 7.6 of the textbook the **bounded solutions of Bessel's equation of order** n on the interval $[0, 1]$ are standardly denoted $J_n(x)$, for $n = 0, 1, 2, \ldots$. In the Maple worksheet $J_n(x)$ is denoted $BesselJ(n, x)$. Please download and open the Maple worksheet Bessel functions in Maple to learn how to use Bessel Functions in Maple. See also the solutions to the exercises in section 7.6 in the textbook to see the Bessel functions at work.

B.6 Quick guide for simple use of Maple

Maple is a symbolic mathematical package that can solve many mathematical problems analytically or numerically with simulations and animations. Maple has friendly user interactive interface. Assume that you have access to one of editions of Maple and are able to open the interactive application, you can get the Maple prompt by clicking the 'File' button, then choose 'New', then choose 'Worksheet Mode'. With the prompt, you can

- Define a function: `f : = x^2 + 1;`

- Differentiate a function with respect to x: `diff(f, x);`

- Get an anti-derivative of a function $f(x)$: `int(f, x);`
  `F:=int(cos(m*x)*sin(n*x), x);`

- Get the value of a definite integral $\int_0^2 f(x)dx$: `int(f, x=0..2);`

- Plot a function $f(x)$ in a interval: `plot(f, x=-1..2);`

- Plot multiple functions in a same graph: `plot([f,x-1,3+x^3], x=-1..2);`

- Define a piecewise function: `f:= piecewise(-2 < x and x < -0.5, 0, -0.5 <x and x< 0.5, 0.5-abs(x), 0.5<x and x<2, 0);`

- Get the Fourier series and plot the partial sum $S_N(x)$.

  ```
  L := 2;
  a_sub_0:= 1/(2*L)*int(f,x=-L..L);
  a_sub_n:= n -> 1/L*int(f*cos(n*Pi*x/L),x=-L..L);
  b_sub_n:= n -> 1/L*int(f*sin(n*Pi*x/L),x=-L..L);

  partsumf:=m->a_sub_0+sum(a_sub_n(n)*cos(n*Pi*x/L)+b_sub_n(n)*sin(n*Pi*x/L),n=1..m);
  partsumf(10);
  plot([f,partsumf(15),partsumf(25)], x=-2*L..2*L,color=[black,red,blue,green]);
  ```

Bibliography

[1] Nakhlé H. Asmar. *Partial Differential Equations with Fourier Series and Boundary Value Problems.* Pearson Prentice Hall, 2000.

[2] John M. Davis. *Introduction to Applied Partial Differential Equations.* W. H. Freeman, 2012.

[3] L. C. Evans. *Partial Differential Equations.* AMS, 1998.

[4] Maple Inc. Maple software package. https://www.maplesoft.com, Latest version, 2020.

[5] Matlab Inc. Matlab software package. https://www.mathworks.com/products/matlab.html, Latest version, 2020.

[6] J. Kevorkian. *Partial Differential Equations.* Wadsworth & Brooks/Cole, 1990.

[7] R. J. LeVeque. *Finite Difference Methods for Ordinary and Partial Differential Equations, Steady State and Time Dependent Problems.* SIAM, 2007.

[8] Z. Li, Z. Qiao, and T. Tang. *An Introduction to Finite Difference and Finite Element Methods for ODE/PDEs of Boundary Value Problems.* Cambridge University Press, 2017.

[9] David J. Logan. *Applied Partial Differential Equations (Undergraduate Texts in Mathematics).* Springer-Verlag, 2014, 3rd Ed.

[10] K. W. Morton and D. F. Mayers. *Numerical Solution of Partial Differential Equations.* Cambridge press, 1995.

[11] Lawrence F. Shampine and Mark W. Reichelt. The MATLAB ODE suite. *SIAM J. Sci. Comput.*, 18(1):1–22, 1997.

[12] Murray Spiegel. *Schaum's Outlines: Laplace Transforms*. McGraw-Hill Education, First edition, 1965.

[13] Murray Spiegel. *Schaum's Outline of Fourier Analysis with Applications to Boundary Value Problems*. McGraw-Hill Education, First edition, 1974.

[14] J. C. Strikwerda. *Finite Difference Scheme and Partial Differential Equations*. Wadsworth & Brooks, 1989.

[15] J. W. Thomas. *Numerical Partial Differential Equations: Finite Difference Methods*. Springer New York, 1995.

Index

1D heat equation, 97
    convolution, 98
    fundamental solution, 98
    solution formula for a Nuemann BC, 101
    solution formula for homogeneous BC, 54
    solution to BVP, 99
    solution to the Cauchy problem, 98
    steady state solution, 102, 103
1D wave equation, 23
    general solution, 24
    left-going wave, 27
    non-homogeneous BC, 92
    right-going wave, 27
    solution formula for homogeneous BC, 50
    solution to Neumann BC, 93
    solution to Robin BC, 96
    with a lower order term, 96
2D Laplace equation
    fundamental solution, 107
2D wave equation
    in polar coordinates, 120
    pole singularity, 120
    separation of variables, 120
    with radial symmetry, 120
3D Laplace equation
    in spherical coordinates, 123
    with radial symmetry, 123

inner product in L^2, 36
weak solution, 10

advection equation, 9, 10
    Cauchy problem, 11, 139
    general solution, 10
    solution to BVP, 14
    solution to Cauchy problem, 11
    variable coefficients, 17
algebraic operations in maple, 176
animate, 184
arithmetic in maple, 176
assignment operator, 178

Bessel equations, 121
    order p, 121
Bessel functions, 121
boundary value problems, BVP, 4
built in functions in Maple, 180

Cauchy problem
    1D wave equation, 25
    advection equation, 11
Cauchy-Schwartz inequality in L^2, 38
classical Fourier Series, 67
classical solution, 10
command line, 172
convergence of Fourier series, 69
convolution
    Fourier transform, 138
    Laplace transform, 143
cosine transform, 134, 144

D'Alembert's formula, 26, 91
diffusion, 140
Dirac delta function, 136
Dirichlet BC, 43

eigenpair, 44
eigenvalue problem, 5, 34
elliptic PDE, 6
Euler's equation, 117, 124
    indicial equation, 117
even extension, 80
execution group, 172
exp, 180
extension and expansion, 77

finite difference method
    1D heat equation of BVPs, 156
    1D wave equation, 155
    2D Poisson equation, 157
    for two-point boundary value problem, 150
finite difference stencil, 151
five-point stencil, 158
floor function, 62
Fourier coefficients, 67
Fourier integral representation, 133, 134
Fourier series, 42, 61
    cosine, 193
    full, 195
    sine, 189, 192
Fourier transform, 135
    Parseval's relation, 138
frequency domain, 67
function, 179
functional space, 36

general solution
    1D wave equation, 24

Gibb's phenomena, 51, 55, 69, 102
gradient operator, 106
grid, 151
grid points, 151

half-range cosine expansion, 80
half-range cosine expansions, 61
half-range expansion, 79
half-range sine expansion, 61, 81
hat function, 65
Heaviside function, 64
Hilbert space, 41
hyperbolic PDE, 6
hyperbolic sine and cosine functions, 108
hyperbolic trig functions, 181

indicial equation, 125
initial value problems, IVP, 4
inner product in L^2, 36
Int, 187
int, 187
inverse Fourier transform, 135
inverse Laplace transform, 143

Laplace equation, 105
    in polar coordinates, 116
    on a circle, 116
Laplace operator, 106
Laplace transform, 141
    convolution, 143
Legendre equation, 125
Legendre polynomial, 198
linear space, 65
ln, 181
local truncation error, 151
Lotka-Volterra predator-prey model, 149

maple list, 176

maple set, 176
master grid point, 158
Matlab ODE Suite, 148
method of separation of variables
    1D heat equation, 53
method of changing variables, 9
    1st order linear non-homogeneous PDEs, 18
    for 1D wave equations, 23
method of characteristic, 12
method of separation of variables, 33

natural ordering, 159
Neumann BC, 43
norm in L^2, 37
normal direction, 106
normal mode solution, 29
normalized orthogonal set, 39
numerical solutions, 2

odd extension, 80
ODE-IVP, 148
orthogonal function in L^2, 36
orthogonal set, 39
    function expansion, 40

parabolic PDE, 6, 97
partial derivative, 186
partial sum, 67, 69, 188
PDE, v
period, 62
periodic boundary condition, 117
periodic function, 62
piecewise continuous function, 64
piecewise smooth function, 65
plot, 182
pointwise convergence, 84, 85
polar coordinates, 114
pole condition, 116

red-black ordering, 159
Robin or mixed BC, 43
round-up, 72, 76

sawtooth function, 63, 64
separation of variables
    wave equations, 91
series solution of 1D heat equation, 54
sine transform, 134, 144
special symbols in Maple, 178
steady state solution, 103, 106
step function, 79, 135
Sturm-Liouville problem, 43
Sturm-Liouville eigenvalue problem, 42
    regular, 45
    singular, 45
sum, 189
superposition, 94
symmetric positive definite, 161

text line, 173
time domain, 67
triangle inequality in L^2, 38
triangular wave, 72
trig functions, 181
trivial solution, 34, 43

uniform Cartesian grid, 151
uniform convergence, 85

Weierstrass M-test theorem, 85
weight function, 43
weighted inner product, 38
worksheet, 172

Lightning Source UK Ltd.
Milton Keynes UK
UKHW050823051021
391695UK00003B/99